13th International Conference on Metrology and Properties of Engineering Surfaces 2011

Journal of Physics: Conference Series Volume 311

Twickenham, United Kingdom
12 – 15 April 2011

ISBN: 978-1-61839-131-5
ISSN: 1742-6588

Printed from e-media with permission by:

Curran Associates, Inc.
57 Morehouse Lane
Red Hook, NY 12571

Some format issues inherent in the e-media version may also appear in this print version.

Printed by Curran Associates, Inc. (2011)

For permission requests, please contact the Institute of Physics
at the address below.

Institute of Physics
Dirac House, Temple Back
Bristol BS1 6BE UK

Phone: 44 1 17 929 7481
Fax: 44 1 17 920 0979

techtracking@iop.org

Additional copies of this publication are available from:

Curran Associates, Inc.
57 Morehouse Lane
Red Hook, NY 12571 USA
Phone: 845-758-0400
Fax: 845-758-2634
Email: curran@proceedings.com
Web: www.proceedings.com

TABLE OF CONTENTS

PAPERS

PREFACE

The 13th International Conference on Metrology and Properties of Engineering Surfaces focused on the progress in surface metrology, surface characterisation instrumentation and properties of engineering surfaces. The conference provided an international forum for academics, industrialists and engineers from different disciplines to meet and exchange their ideas, results and latest research.

The conference was held at Twickenham Stadium, situated approximately six miles from Heathrow Airport and approximately three miles from the *National Physical Laboratory (NPL)*.

This was the thirteenth in the very successful series of conferences, which have firmly established surface topography as a new and exciting interdisciplinary field of scientific and technological studies.

Scientific Themes:

- Surface, Micro and Nano Metrology
- Measurement and Instrumentation
- Metrology for MST Devices
- Freeform Surface Measurement and Characterisation
- Uncertainty, Traceability and Calibration
- AFM/SPM Metrology
- Tribology and Wear Phenomena
- Functional Applications
- Stylus and Optical Instruments

Journal of Physics: Conference Series **311** (2011) 012001 doi:10.1088/1742-6596/311/1/012001

Fast algorithm for Morphological Filters

Shan Lou, Xiangqian Jiang, Paul J. Scott

Centre for Precision Technologies, University of Huddersfield, Huddersfield HD1 3DH, UK

E-mail: x.jiang@hud.ac.uk (Xiangqian Jiang)

Abstract. In surface metrology, morphological filters, which evolved from the envelope filtering system (E-system) work well for functional prediction of surface finish in the analysis of surfaces in contact. The naive algorithms are time consuming, especially for areal data, and not generally adopted in real practice. A fast algorithm is proposed based on the alpha shape. The hull obtained by rolling the alpha ball is equivalent to the morphological opening/closing in theory. The algorithm depends on Delaunay triangulation with time complexity O(nlogn). In comparison to the naive algorithms it generates the opening and closing envelope without combining dilation and erosion. Edge distortion is corrected by reflective padding for open profiles/surfaces. Spikes in the sample data are detected and points interpolated to prevent singularities. The proposed algorithm works well both for morphological profile and area filters. Examples are presented to demonstrate the validity and superiority on efficiency of this algorithm over the naive algorithm.

1. Introduction

Surface metrology is the measurement of small scale geometrical features on surfaces. It has profound influences on manufacturing quality as it plays two important roles: on the hand it helps to control the manufacturing process; On the other hand it helps functional prediction. In surface metrology, the extraction and separation of multi-scale features mainly depend on filtering techniques. Among various filtering techniques, morphological filter is most suitable for functional prediction as its logic is more related with the geometrical properties of surfaces.

The envelope filtering system, called the E-System, is archived by rolling a ball (resp. disk) over the surface (resp. profile) [1]. By introducing morphological operations into the E-system [2], morphological filters emerged. They are carried out by performing morphological operations (dilation, erosion, opening, and closing) on the input profile/surface with the circular or flat structuring element, usually circular. Over the decade morphological filters have found many practical applications, for instance, the approximation of the conformable interface of two mating surface [3], the analysis of liner surface comprised of deep valleys and plateau roughness [4], etc. Although morphological filters could give better results in functional prediction and be regarded as the complement of mean-line filters, they are not generally adopted in real practice. One of the most important reasons is it is time consuming, especially for areal data. A naive algorithm was described in ISO 16610 [5]. It places the structuring element on each sample point and calculates the extreme points. The naive algorithm runs slow, especially for large surface and large structuring element. Meanwhile it combines the dilation and erosion to generate the opening and closing, doubling the processing time. In addition it is limited to the uniform sampled profile/surface. This paper sets out to develop a faster and more general algorithm for morphological filters. The algorithm employs the alpha shape as the basis of computation.

2. Introduction to Alpha Shape

The alpha shape was introduced by Edulsbrunner in the 1980's [6]. The alpha hull is obtained by rolling the ball with radius α over the point set. Straightening the round faces of the alpha hull by line segments for arcs and triangles for caps generates the alpha shape. The computation of the alpha shape is based on the Delaunay triangulation. For a point set S, the boundary of its alpha shape is equivalent to the boundary of the alpha complex C_α, which is the collection of the simplices in the Delaunay triangulation satisfying two properties: (i) the radius of the smallest circumsphere of the simplex is smaller than α and is empty; (ii) the simplex is a face of super simplex in the alpha complex.

Worrings and Smedulers [7] proved that the alpha hull is equivalent to the closing of the point set X with a generalized ball of radius $-1/\alpha$ and that from the duality of the closing and the opening the alpha hull is the complement of the opening of X^c (complement of X) with the same ball as the structuring element. This crucial relationship provides a feasible method to compute morphological filters by the alpha hull.

3. Fast Algorithm for Morphological Filters

3.1. End Distortion Correction

End distortion is common for the filtration of open profiles/surfaces. Morphological filters are no exception. For mean-line filters, a common solution is to add sufficient zeros to the beginning and end of the profile/surface, referred as zero-padding. Zero-padding is not suitable for morphological filters because the padded part of the surface should not be geometrically viewed as a flat plane with zero height. Instead it should reflect the geometrical features of the surface edge. To achieve this, the padding of the surface is conducted by reflecting the counterpart area of the surface in the range of the ball radius starting from the surface margin. Figure 1 demonstrates the reflective padding of the surface.

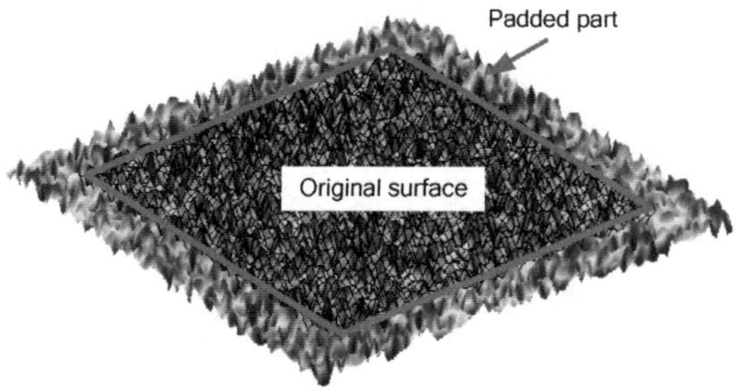

Figure 1. Padding the surface by refection

3.2. Spikes Detection and Points Interpolation

The sample data is a discrete representation of the measured surface. It may happen that sharp spikes exist in the sample data. Sometimes the space between the peak point and pit point is a lot larger than the rolling ball which will run into the interior of the surface. It is not allowed in reality because the real surface is physically continuous and won't allow the ball to center. Therefore sharp spikes should be detected and more points linearly interpolated on the ridge of the spike to prevent the ball from passing through. Figure 2 illustrates the whole process. P_1, P_2, ..., P_{10} are the sample points from the original profile. $P_3P_4P_5$ form a local peak. P_3 and P_4 are spaced far from each other so that the ball will roll into the interior. In this case additional points I_1, I_2 (and more if needed) are linearly interpolated to reduce the gap between P_3 and P_4. For surfaces, they can degenerate to profiles if they are considered as the composition of parallel profile sections. There is a trivial difference between the closing envelope and the opening envelope in their spike detection. For the closing envelope, it suffices to detect peak spikes only in that the closing envelope is only determined by peaks, and valleys can be ignored. As opposed for the opening envelopes it is enough to search valleys because it is only affected by valleys.

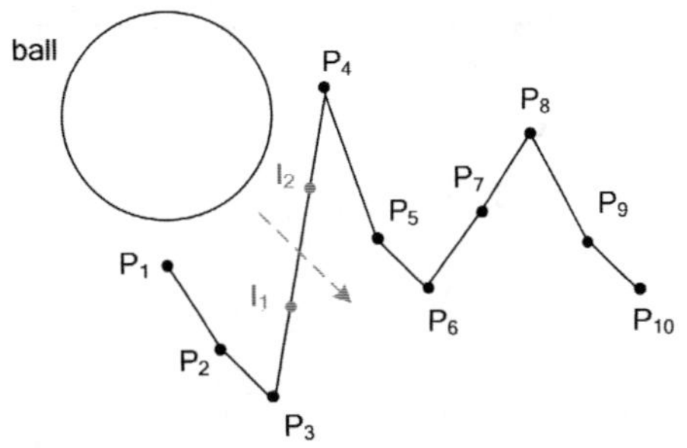

Figure 2. Spikes detection and points interpolation in sample data.

3.3. Boundary Extraction

Following the reflective padding, spikes detection and point interpolation, the next step is to triangulate the sample points by the Delaunay triangulation and subsequently extract the boundaries of the alpha shape ∂S_α, i.e. the boundaries of the alpha complex ∂C_α. The three-dimensional Delaunay triangulation results in a series of tetrahedrons, which could be categorized into two groups: tetrahedrons T_p whose circumscribe sphere radius is smaller than the radius of the rolling ball α, and tetrahedrons T_{np} whose circumscribe sphere radius is equal or larger than α. By the definition of the alpha complex, T_p is contained in C_α.

T_p consists of the interior triangles σ_{int} and the triangles σ_{reg} that bound T_p. σ_{reg}, called the regular boundaries, are a part of ∂C_α and they bound super simplices (i.e. tetrahedrons) in C_α. T_{np} is comprised of three components: the triangles σ_{ext} that are out of C_α, part of σ_{reg} that are shared by both T_p and T_{np}, and the triangles σ_{sing} that are the other part of ∂C_α. σ_{sing}, called the singular boundaries, differ from σ_{reg} in that they do not bound super simplices in C_α. σ_{sing} satisfy two conditions: (i) the radius of its smallest circumsphere is smaller than α; (ii) its smallest circumsphere is empty. The regular boundaries σ_{reg} and the singular boundaries σ_{sing} make up all the boundaries of C_α. Figure 3 illustrates the boundary triangles of the alpha shape of sample points on a surface.

3.4. Envelope Calculation

The final step is to compute the envelope points. For each sample point, there is a corresponding point on the envelope locating at the same sample position. These points form a discrete representation of the envelope. To obtain the envelope point, an infinite vertical line is placed at the sample position to intersect with the covering boundary triangles (See figure 4). For the closing envelope, find the triangle with the highest intersection point. Then the envelope point can be obtained by intersecting the vertical line with the circumsphere of this triangle. The lower intersection point is the envelope point. Likewise, the envelope point for the opening envelope can be calculated in similar way.

Journal of Physics: Conference Series **311** (2011) 012001 doi:10.1088/1742-6596/311/1/012001

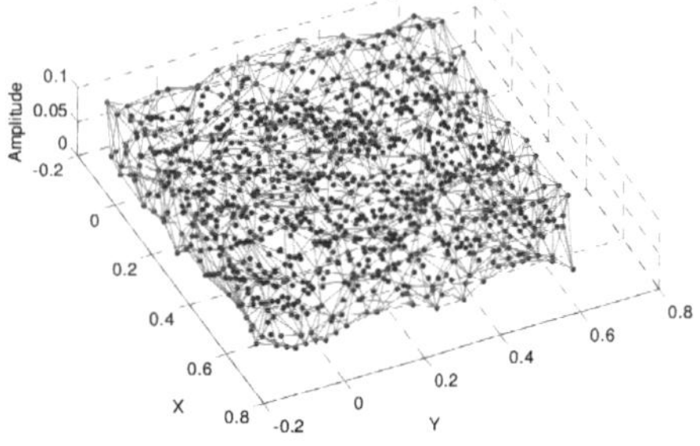

Figure 3. Boundary triangles of the alpha shape of measured.

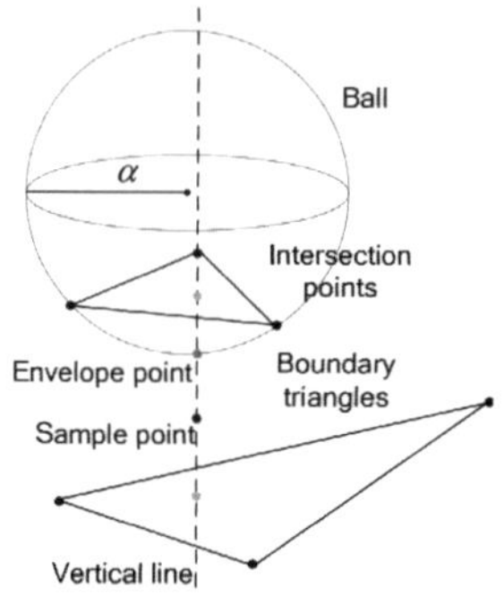

Figure 4. Calculation of the envelope point

4. Examples

Examples of the closing envelope filter are shown in figure 5-8. Figure 5 presents a surface with 250*250 sample points. The total sampling area is 1.25*1.25 mm with sampling internal 5 μm. The structuring element is a 5 mm ball. Figure 6 illustrates the closing envelope superimposed over the surface. Figure 7 shows a sectional profile of this surface, its corresponding closing profile envelope is presented in figure 8. Experiment data shows the proposed algorithm is more efficient than the naive algorithm when processing with large areal data and large structuring elements.

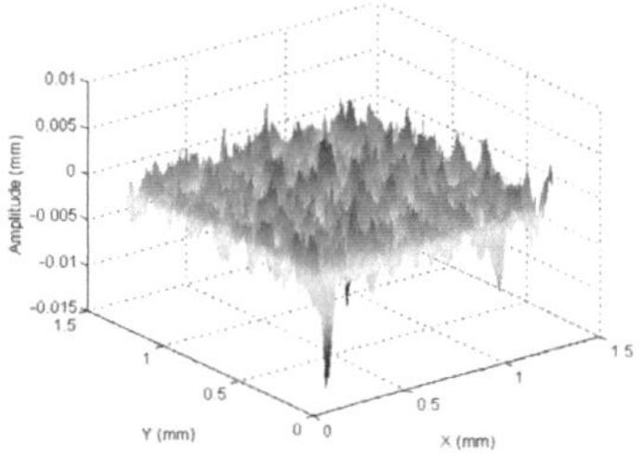

Figure 5. Surface: 250*250 sample points.

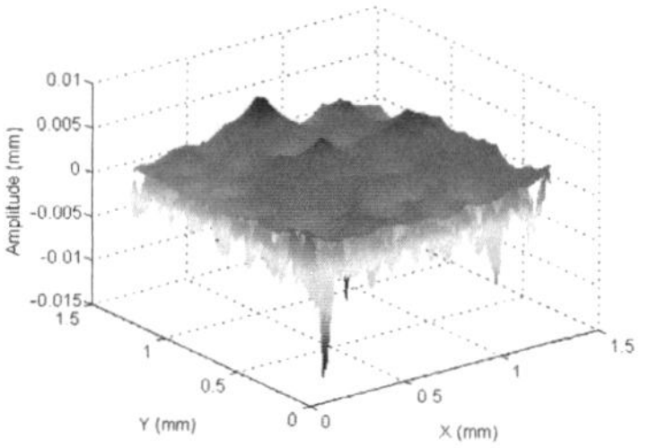

Figure 6. Closing envelope superimposed over surface.

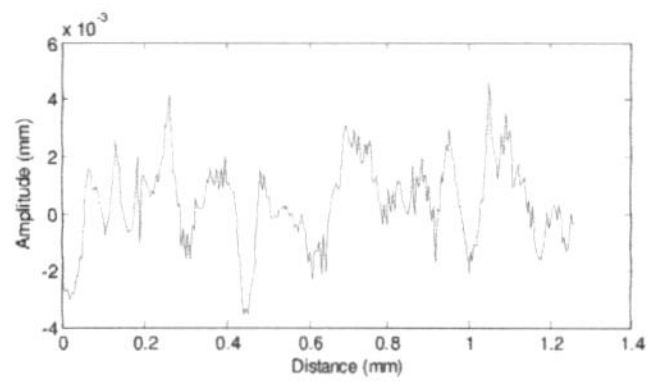

Figure 6. Profile: 250 sample points.

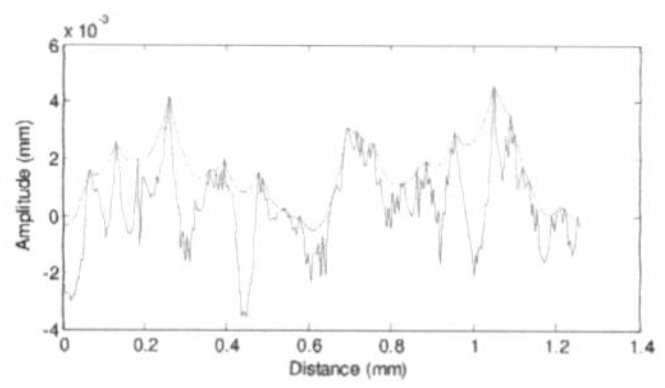

Figure 7. Closing envelope superimposed over profile.

5. Conclusion

The relationship that the alpha hull of the point set is theoretically identical to its closing/opening provides a feasible approach for computing morphological filters with the circular structuring elements. The proposed algorithm employs the alpha shape as the basis of computation. The algorithm depends on Delaunay triangulation with time complexity *O(nlogn)*. It directly yields the closing/opening envelope without combining the dilation and erosion operations. Prior to calculating the alpha shape, special processes are conducted on the sample data to correct end distortion and justify the singularities caused by spikes. The proposed algorithm works well for both morphological profile and area filters. It is more efficient than the naive algorithm in processing speed for large areal data and large structuring element. Moreover, the proposed algorithm works for non-uniform sampled profile/surface, being more general than the naive algorithm. Another notable merit of the proposed algorithm is that the boundary triangulation data can be reused for multiple attempts of the ball radius. It could save a great deal of computing time considering in real practice a multitude of trails may be made for an appropriate ball radius.

Acknowledgements
The authors would like to thank the University of Huddersfield for providing the scholarship to support this work.

References

[1] Von Weingraber H 1956 Zur Definition der Oberflächenrauheit *Werkstattstechnik Masch Bass* **46**
[2] Srinivasan V 1998 Discrete morphological filters for metrology, *Proc. 6th ISMQC Symposium on Metrology for Quality Control in Production*
[3] Malburg C M 2003 Surface Profile Analysis for Conformable Interfaces *Transactions of ASME.* **125** 624-627
[4] Decenciere E and Jeulin D 2001 Morphological decomposition of the surface topography of an internal combustion engine cylinder to characterize wear *Wear* **249** 482-488
[5] ISO 16610-41 2005 Geometrical Product Specification (GPS)-Filtration Part 41: Morphological profile filters Disk and horizontal line-segment filters
[6] Edelsbrunner H and Mucke E P 1994 Three-dimensional alpha shapes *ACM Trans. Graph.* **13** 43-72
[7] Worring M and Smelders W M 1994 Shape of arbitrary finite point set in R^2 *Journal of Mathematical Image and Vision* **4** 151-170

Realization and calibration of the "Isara 400" ultra-precision CMM

I Widdershoven, R L Donker, H A M Spaan

IBS Precision Engineering, Esp 201, 5633AD, Eindhoven, The Netherlands

E-mail: widdershoven@ibspe.com

Abstract. This paper presents the realization of the Isara 400 ultra-precision 3D coordinate measuring machine, which features a measuring volume of 400 x 400 x 100 mm and a volumetric (3D) measurement uncertainty of 100 nm (2σ). In order to achieve these challenging specifications, specific calibration strategies need to be applied, such as the calibration of flatness and out-of-squareness of the system's mirror table. In addition, a newly developed ultra-precision tactile probe system is described, featuring a probe tip with a diameter of 70 µm; results of the 3D sensitivity calibration of this probe are presented.

1. Introduction

The Isara 400 coordinate measuring machine (CMM) is the latest development of IBS Precision Engineering for coordinate metrology of large, complex parts with nanometer level measuring uncertainty. The expected 3D measuring uncertainty is 100 nm (2σ) within the complete measuring volume of 400 x 400 x 100 mm. The Isara 400 CMM is capable of measuring complex surfaces like aspheres, free-forms or integrated optics with nanometre accuracy in full 3D, and thus overcomes some of the limitations presented by the optical methods or contacting profilers which are currently used. In addition, application areas include the geometrical inspection of a wide range of industrial parts, similar to conventional CMMs, but with much higher measuring accuracy.

This paper presents the realized design as well as the most critical calibration strategies necessary to achieve the challenging accuracy specifications.

2. Design

The Isara 400 CMM features a measurement volume of 400 x 400 x 100 mm. Three plane mirror laser interferometers are applied as measuring systems for the machine axes. The interferometers each measure against the sides of a mirror table, on which the work piece is mounted. These interferometers are mounted in a single body metrology frame, which also holds the probe system (see figure 1a). The laser beams are aligned to the probe tip and their mutual alignment does not change during movement of the axes, thus fulfilling the Abbe principle [2] in 3D within the complete measuring volume. As a result, straightness errors and rotations of the three translation stages will have no first order influence on the measurement result. The influence of flatness and squareness errors of the three mirrors is eliminated by means of a series of on-machine calibration measurements.

In the configuration of the translation axes, a machine concept was chosen in which the variable product mass does not need to be moved vertically. The product is mounted on the mirror table, which moves only in X- and Y-direction over a granite base plate, guided by air bearings in a 'floating table' configuration. The complete metrology frame moves in Z-direction, with guiding provided by air bearings against a vertical granite surface (figure 1a). Figure 1b shows an overview of the complete machine.

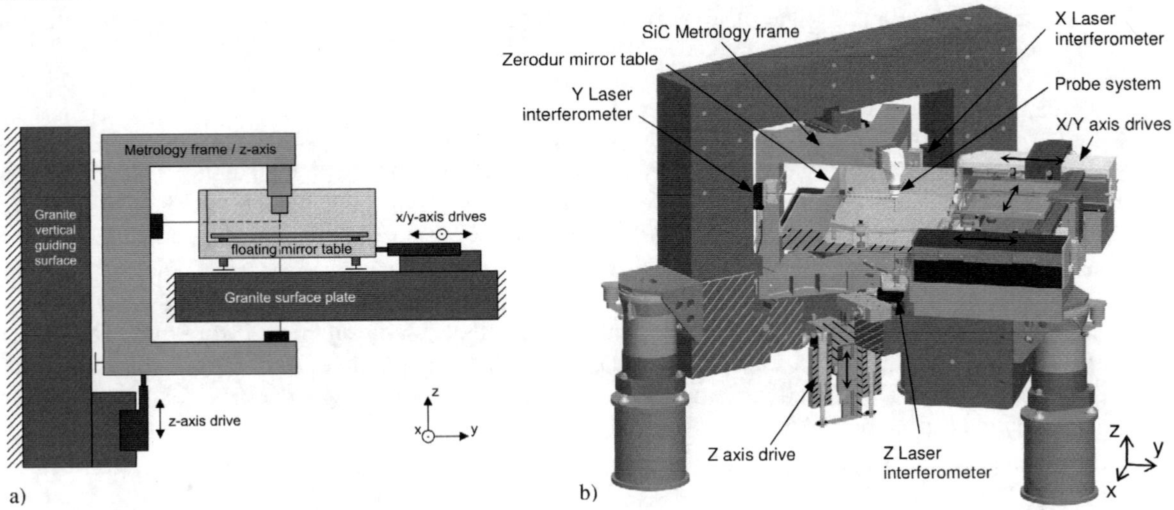

a) b)

Figure 1. Design overview of the Isara 400 CMM; a) concept sketch; b) full 3D CAD design (without covers)

The function of the metrology frame, shown in figure 3a, is to maintain the mutual position and alignment of the probe and the three laser interferometers with high stability. The metrology frame was designed as an assembly of hollow beams of silicon carbide (SiC), resulting in a structure which is both stiff and lightweight, while also providing good thermal stability.

The mirror table of the Isara 400 is a monolithic Zerodur part with three reflective sides. The mirror table is supported by three flat air bearings, whose preload is provided by the weight of the mirror table assembly with product. Figure 2b shows the mirror table and the X/Y-drives; for more details see [3]. Work pieces are not placed directly onto the Zerodur, but onto a removable SiC product table, which serves as an interface between product and mirror table. The weight of the product table with work piece is directly transferred through its mounting supports to the supporting air bearings, without causing additional deformation of the Zerodur mirror table.

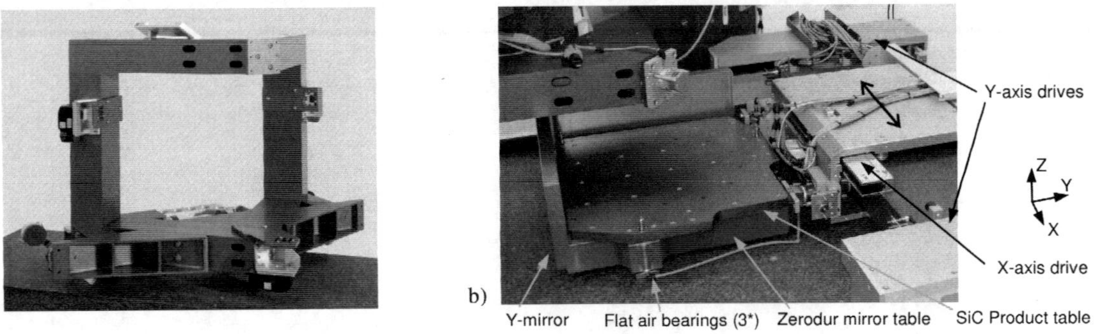

a) b)

Figure 2. a) Silicon carbide metrology frame; b) Mirror table and X/Y drives.

3. Mirror table calibration

The three plane mirrors of the mirror table serve as targets for the laser interferometers and can thus introduce measuring errors due to flatness deviations and out-of-squareness of the mirrors. These deviations need to be calibrated and compensated to fulfill the high accuracy demands. All mirror table calibration measurements are performed on the machine itself, in its final assembled position, thus the sagging of the mirrors due to gravity is included. The calibration strategies presented here build on the work described in [4], but effort has been made to simplify the procedure, for example by making use of the flatness data of a calibrated flatness reference artefact in stead of performing an elaborate series of reversal measurements with an artefact whose flatness is unknown.

3.1. Flatness calibration

The flatness of the mirror table is calibrated by placing a calibrated flatness reference in the measurement volume of the machine and performing a flatness measurement with a highly accurate capacitance probe. The reference artefact in this case is a Zerodur block, which has three sides with a reflective metal coating. Figure 3a shows a sketch of the setup for flatness calibrating of the Z-mirror. The complete top surface of the artefact is measured with the capacitance probe; the measurement result is the sum of the flatness deviations of the mirror and the reference artefact. As the flatness from the reference artefact is known from optical calibrations, this can be subtracted from the measurement result, leaving only the flatness deviation of the Z-mirror, which is then used to correct the measured displacements from the laser interferometers. A similar calibration is performed for the X- and Y-mirrors, by measuring the sides of the same artefact.

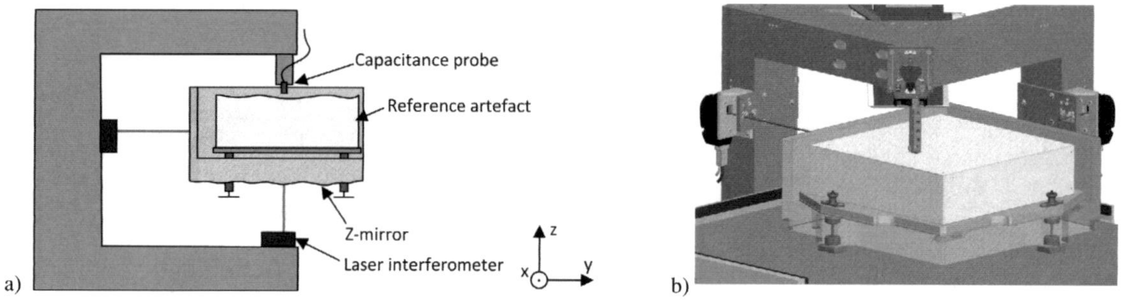

Figure 3. Flatness calibration of the Z-mirror; a) concept sketch; b) complete setup

3.2. Optical flatness calibration of reference artefact

A critical part of this calibration strategy is the calibration of the flatness reference. The Zerodur reference artefact is calibrated using Fizeau interferometry. This optical calibration provides the flatness mappings shown in figure 4, which can be subtracted from the flatness measurement described in the previous paragraph.

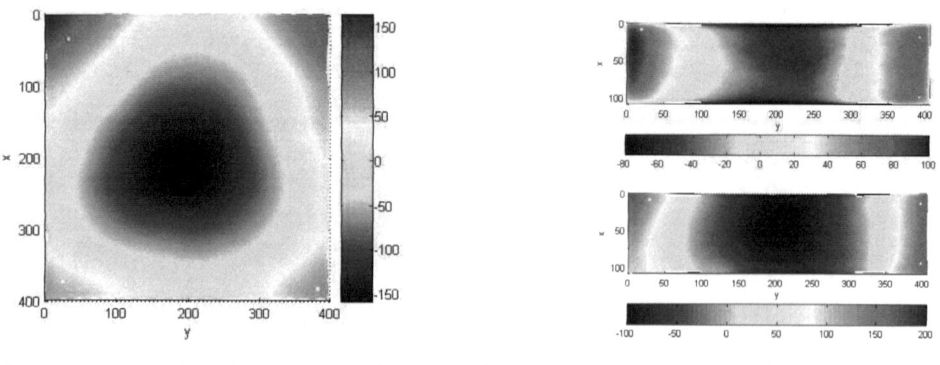

Figure 4. Results of the optical flatness calibration of the three sides of the artefact; colour scales in nanometres.

In each optical calibration measurement, the artefact is in the same orientation as it is during the calibration on the machine (figure 3b), so no additional sagging needs to be taken into account. The measurement uncertainty of the optical flatness calibration is less than 10 nm (k=2).

3.3. Out-of-squareness calibration

The orthogonality of the machine's metrology coordinate system is determined by the orthogonality of the three mirrors of the mirror table. Any out-of-squareness between these mirrors will cause measurement errors. Using the same Zerodur artefact as described in the previous paragraph, the out-of-squareness of the three mirrors can be calibrated. The out-of-squareness of the artefact has not been accurately calibrated; error separation needs to be applied to distinguish between the out-of-squareness of the mirror table and that of the artefact.

All three coated sides of the artefact are measured by the capacitance probe, without changing the orientation of the artefact with respect to the mirror table. The mutual angle between the three measured planes is determined; this measurement result is a combination of the out-of-squareness of the mirror table and the out-of-squareness of the artefact. By measuring the artefact in multiple orientations, it is possible to perform error separation.

The procedure for one specific out-of-squareness angle is shown in figure 5. The mounting orientation of the capacitance probe varies in order to measure several sides of the artefact. In the first orientation of the artefact, the measured angle between the two planes equals $M1=\varphi-\alpha YZ+90°$, where φ and αYZ are the out-of-squarenesses of the artefact and the mirror table, respectively. In the second orientation, the measured angle is $M2=\varphi+\alpha YZ-90°$. Because the contribution of αYZ changes sign between the two measurement results, it is possible to determine both φ and αYZ from these two measurements: $\alpha YZ=½\cdot(180-M1+M2)$ and $\varphi=½\cdot(M1+M2)$. A similar strategy is used for the other out-of-squareness errors.

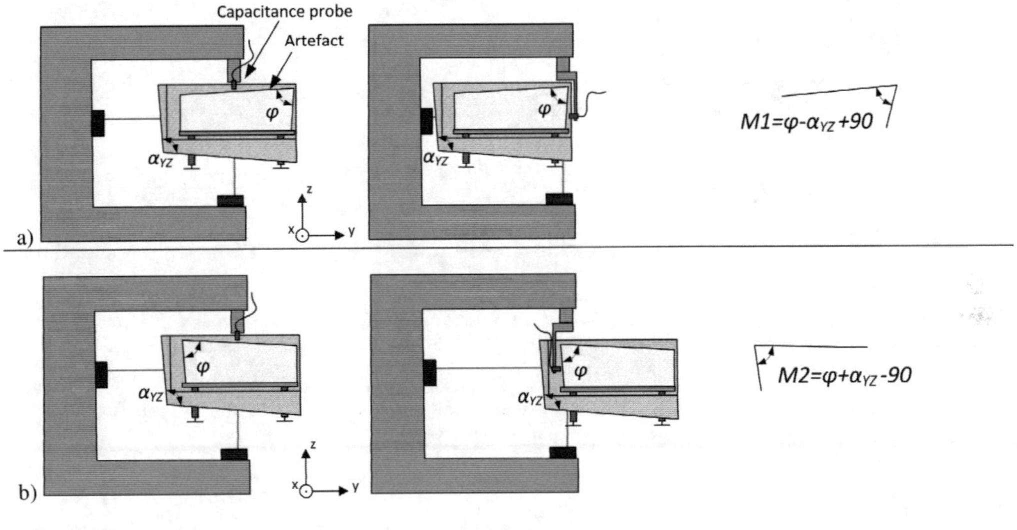

Figure 5. Out-of-squareness calibration of Y- and Z-mirrors; a) artefact orientation 1; b) artefact orientation 2.

4. Tactile probe calibration

The design and calibration of the newly developed "Triskelion" probe system is described in detail in [1]. The design features an elastically suspended stylus, which is free to deflect in X-, Y- and Z-direction at its tip; this deflection is measured by three capacitance sensors which are integrated in the probe system.

A miniaturized version of this probe system features a stylus with a tip diameter of about 70 µm (see figure 6a). The small tip enables measurements of very small features, such as the inside diameter of very small holes (up to 1 mm depth). The elastic suspension of this miniaturized probe system was also redesigned to further reduce probing forces, to prevent work piece surface damage. Figure 6b shows a photograph of the realized probe.

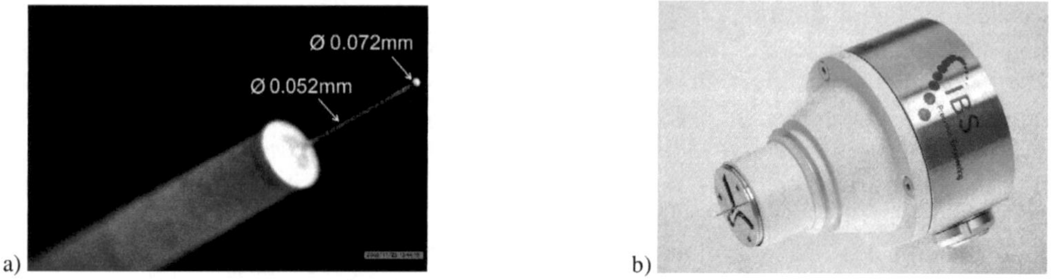

Figure 6: a) Photograph of miniaturized stylus. b) Realized miniature Triskelion probe

The sensitivity calibration of this probe system is performed on an ultra-precision CMM. The probe is placed in contact with a flat work piece surface, which is located on the product table. Probe deflection is then applied by moving the table; the output signals of the probe and the interferometric table displacement are logged. Repeating this measurement for multiple probing directions yields a 3D sensitivity model. This model is validated by performing additional probing measurements. One such result is presented in figure 7 (unfiltered data). For probe tip deflections ≤ 5 μm, measurement errors are < 10 nm per axis of the coordinate system and < 15 nm in 3D.

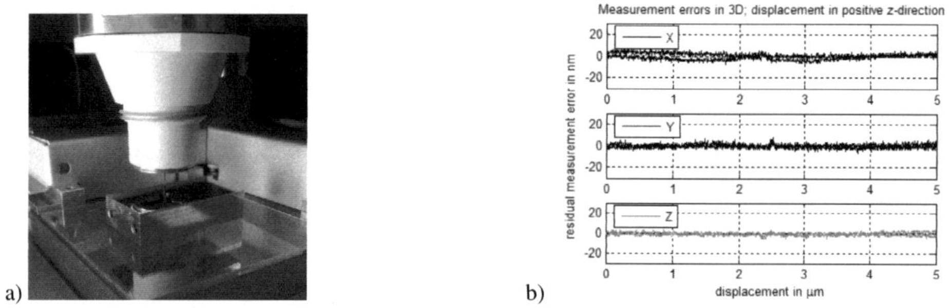

Figure 7: a) Miniature probe in CMM. b) Residual measurement errors (x,y,z) for probing in z-direction

5. Conclusions

The Isara 400 ultra-precision CMM has been realized and is currently operational at IBS Precision Engineering. Calibration of the mirror table deviations are critical to achieving the targeted volumetric uncertainty of 100 nm (k=2). A newly developed, miniaturized tactile probe system has been realized and the sensitivity calibration has been successfully performed.

Acknowledgements

The work presented in this document is part of the European sixth framework programme projects NanoCMM, FP6-026717-2 and Production4μ, FP6-2004-NI-4.

The authors wish to thank ZYGO Corporation for their contribution regarding the optical calibration of the reference artefact.

References

[1] H. A. M. Spaan, R. L. Donker, and I. Widdershoven, Isara 400: Development of an ultra-precision CMM for 3D measurement of large parts, *Proc. ASPE spring meeting*, 2009.
[2] E. Abbe, Meßapparate für Physiker, Zeitschrift Für Instrumentenkunde, 1890; 10: 446-448.
[3] R. Donker, I. Widdershoven, D. Brouns, H. Spaan, Realization of Isara 400: a large measurement volume ultra-precision CMM, *Proc. ASPE annual meeting*, 2009.
[4] T.A.M. Ruijl, Ultra Precision Coordinate Measuring Machine – Design, Calibration and Error Compensation, PhD thesis, Delft University of Technology, The Netherlands, 2001, ISBN 90-6464-287-7

Journal of Physics: Conference Series **311** (2011) 012003 doi:10.1088/1742-6596/311/1/012003

Optical Measurement of Micro Cutting Tools

F Helmli, R Danzl[1] and S Scherer
Alicona Imaging GmbH,
Teslastraße 8, 8074 Grambach, Austria

E-mail: reinhard.danzl@alicona.com

Abstract. We present an optical measurement device for the complete 3D measurement of very small micro cutting tools with diameters down to 100µm and below. Such micro tools can be hardly measured by conventional devices such as tactile systems due to their complex geometry and small dimensions. The used optical device is based on the Focus-Variation principle which combines the small depth of focus of an optical system with vertical scanning to provide topographical and color information from the variation of focus. Full 360° measurements of the tools are automatically measured in a two step procedure. First, single measurements are performed from different directions all around the specimen with the aid of a motorized rotation unit. Then all single datasets are automatically aligned and merged to a full 360° dataset. Various measurement results are presented for milling cutters and drills including tools with diameters in the range of about 50µm. From the measured tools various parameters can be obtained including the major diameter, the core diameter or various angles. Additionally roughness measurements are provided on small cutting tools showing the versatility of the system for both form and roughness measurements.

1. Introduction

The 3-dimensional measurement of cutting tools, drills and milling cutters with diameters in the micro range (e.g. smaller than a few hundred microns) has become an essential aspect in quality assurance. Form and wear of a cutting edge have a strong impact on the manufacturing quality. The dimensional accuracy significantly influences the manufacturing speed as well as the lifetime of a tool. Due to the highly complex geometries (e.g. steep flanks, relief…) the 3D-measurement has become a big challenge for optical measurement devices.

Traditionally, tool measurements are merely performed by tactile measurement devices or with the analysis of images using back light processes. One disadvantage of tactile systems is the long measurement time. Another one is the fact that the geometry of the stylus tip falsifies the measurement result. On top of that, wear of the tactile tip and wear of the measured component are undesired side effects. Moreover, for very small micro cutting tools, tactile systems are not applicable at all. The evaluation and analysis of 2D-images is only possible if no significant areas are hidden in the images. This back-light method is not possible if real 3D-information is required. This applies to tools with concave areas and a variety of parameters of complex tools such as drills or milling cutters.

A typical example for a measurement device especially designed for the measurement of tools is the Helicheck device [1]. This is equipped with a tactile probe and up to four cameras for optical 2D measurements from different directions. Additionally it can be extended by a module for the

[1] To whom any correspondence should be addressed.

measurement of the cutting edge roundness. However, this device does not deliver dense 3D measurements of the whole tool and has limitations when it comes to the measurement of very complex tool parameters. In contrast to such devices that are optimized for tool measurements and also deliver tool parameters in addition to 3D data, miniature coordinate measurement machines (CMM) are available, that are able to measure arbitrary samples (not only tools) from different directions. An example for such CMMs is the F25 device [2] from Carl Zeiss, which consists of a tactile probe and additional cameras for optical 2D measurements. Although this device has a very high accuracy the tactile measurement of very small tools can be problematic due to the above mentioned limitations of tactile probes.

Within this paper a new system for 360° measurement of complex tools is introduced. The InfiniteFocus system (Figure 1) is based on the technology of Focus-Variation [3] [4] and is equipped with a motorized rotation unit. In the first step various measurements are performed at several angles. In step two the single measurements are automatically aligned to each other. As a result a full 360° dataset is available for all further measurements. Geometric primitives can as well be measured as it is possible to analyze contours. Most measurement systems that were available up to now have the decisive disadvantage that they can only measure rather simple parameters as for example the major diameter. For complex parameters a highly complex and time consuming measurement process must be performed. The presented system in contrast does not only measure the full 3D-dataset in high resolution but also delivers complex tool parameters [5]. Parameters such as flank diameter, chipping angle or thread diameter are easily measured. Additionally a variety of roughness parameters can be measured.

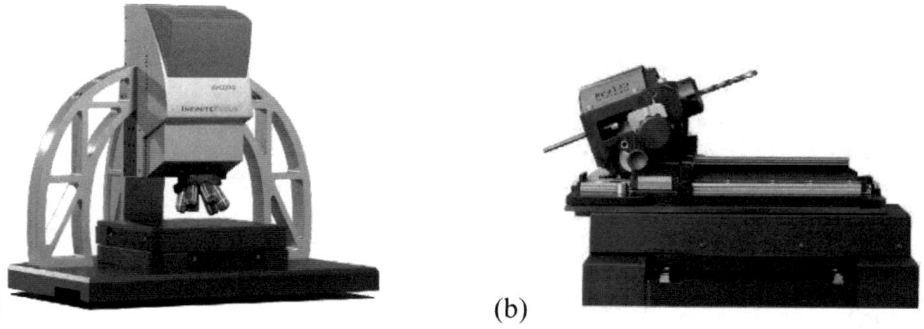

(a) (b)

Figure 1. (a) InfiniteFocus – optical 3D micro coordinate measurement system (b) Rotation and tilting unit for rotating the specimen and performing measurements of 360°.

2. 360° Measurement of Complex Components

The optical 3D-measurements, shown in this paper, were measured with the measurement device InfiniteFocus from Alicona This system is based on the technology of Focus-Variation. This technology especially well handles steep flank measurements. Exactly this ability is necessary for measuring micro drills, micro components or tap tools. The technique of Focus-Variation [3 [4] combines the small depth of focus of an optical system with vertical scanning to provide topographical and color information from the variation of focus. In order to perform 3D measurements the device vertically scans the specimen by the sensor while continuously acquiring data. Since the system has a limited depth of field only small parts of the object are imaged sharply at the same time. By analyzing the variation of focus during the scanning process a 3D-dataset is obtained. The standard system is able to measure surfaces from one direction and to deliver a height value for each x/y position in the field of view. For measuring a tool around 360° a special rotation and tilting unit (Figure 1b) has been developed. This unit allows tilting the tool between 0° and 90° and an endless rotation around 360°. The rotation unit includes a special clamping fixture which allows highly accurate positioning of the tool. The number of required measurements is calculated automatically and is dependent on several parameters. Important factors are the size of the used field of view, the diameter of the tool, the tilting and the scan height. In a second step the alignment is performed in order to receive a highly accurate 360° dataset and to reduce the small but existent error of the rotation axis calibration. In step three all single measurements are combined with a special algorithm resulting in a 360° dataset.

3. Results

In the following examples of drills and milling tools show the scope of the measurement device. In Section 3.1 a drill bit with a major diameter of 260µm is presented, whereas Section 3.2 contains measurements of a micro milling cutter with a diameter of around 70µm.

3.1. Complex Geometry Measurement of a Drill

In Figure 2 the measured micro drill is displayed in pseudo color. The colors represent the distance of every measurement point to the tools' axis. This drill has been measured with the previously presented measurement device InfiniteFocus and its rotation unit. A cutting layer is displayed in Figure 3. The cross section of this dataset is given in Figure 4. By using the contour measurement the major radius of 128.8µm (circle 1) and the core radius of 59.2µm (circle 2) were measured. Since the dimensional accuracy of a drill has a big influence on manufacturing quality it is important that the drill is produced exactly the way it is thought to be. The measured radius of circle 3 is 97.1µm and those of circle 4 97.4µm. There is only a difference of 0.3µm.

In order to evaluate the accuracy of the measurements the core and major radius has been measured at seven different axial positions in a range of 300µm as shown in Table 2. An axial position of 500µm corresponds to the cross section shown in Figure 3. The tip position corresponds to an axial position of 1050µm. The major radius stays merely constant over the whole range whereas the core radius linearly decreases towards the tip position.

For the major radius a standard deviation of 0.192µm has been measured which is very small for a micro drill measurement. In order to evaluate a useful standard deviation for the core radius a regression line has been fitted into the measured core radius values and the deviation to this regression line has been calculated. The standard deviation of these deviations is 0.19µm and thus very similar to those of the major radius. The trend of the major and the core radius is visualized in Figure 5.

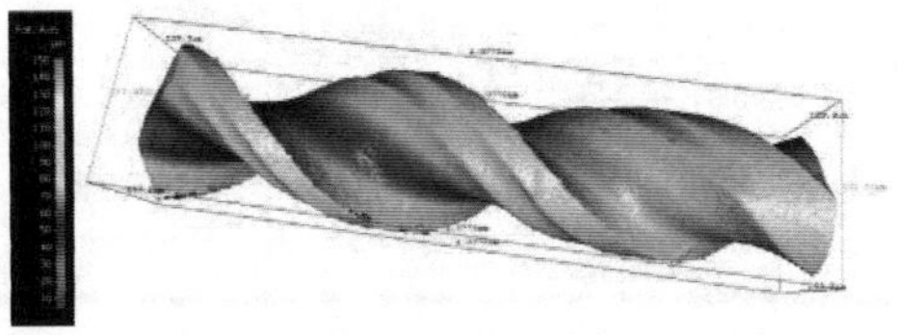

Figure 2. 360° Measurement of a micro drill with major radius of about 130µm. The bounding box is 0.26 x 0.26 x 1.08mm.

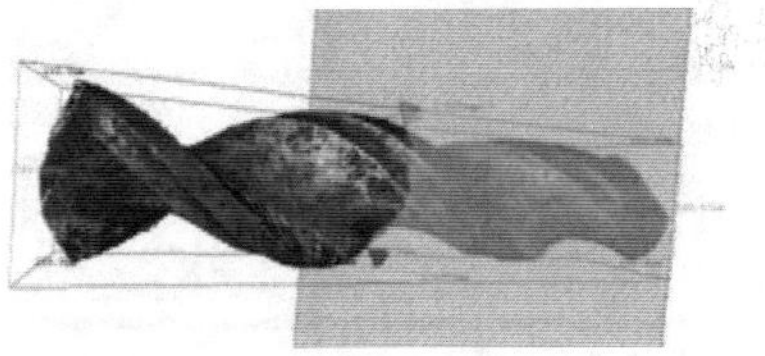

Figure 3. Measured micro drill with extracted cutting layer for contour measurement. The bounding box is 0.26 x 0.26 x 1.08mm.

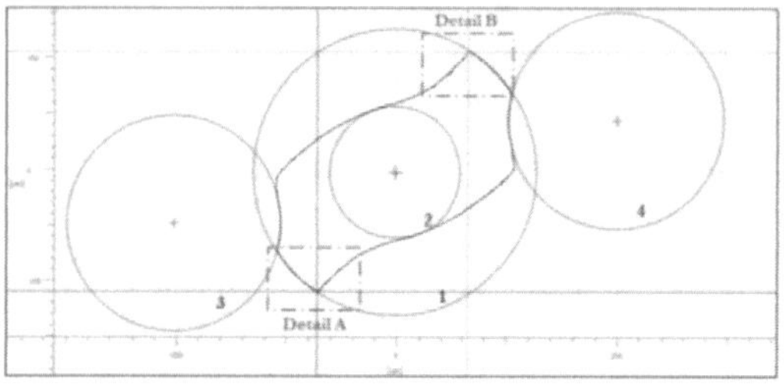

Figure 4. Contour measurement including several circle measurements and showing the areas for the cutting edge detail A and B. The coordinate center (0/0) is the center of circle 2. The y-scale ranges from -200 to +150µm the x-scale from -350 to +350µm.

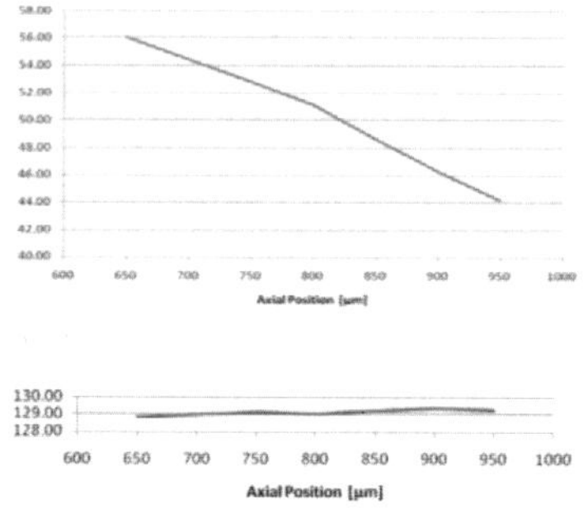

Figure 5. Upper diagram: Core radius at different axial positions; Lower diagram: Major radius at different axial positions.

Table 1. Measurement of the major and core radius at different axial positions (all values are in µm).

Number	Axial Position	Major Radius	Core Radius	Regression Line	Deviation to Regression Line
1	650	128.80	56.03	56.48	0.45
2	700	128.95	54.40	54.48	0.08
3	750	129.10	52.78	52.48	0.29
4	800	129.01	51.12	50.48	0.63
5	850	129.20	48.61	48.49	0.12
6	900	129.36	46.26	46.49	0.23
7	950	129.25	44.20	44.49	0.29
Standard Deviation		0.192			0.190

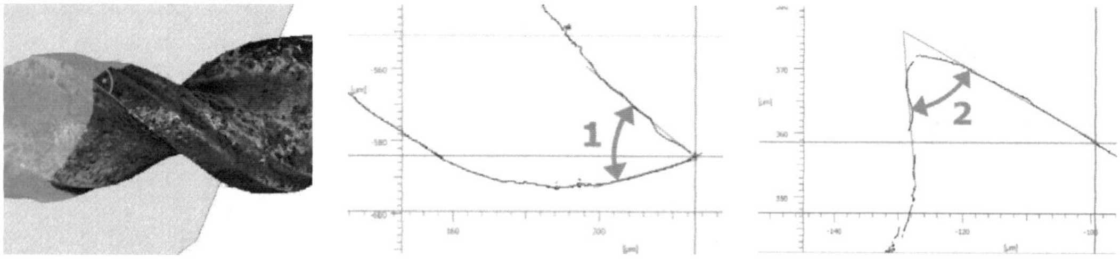

(a) (b) (c)

Figure 7. (a) Section of Detail A. (b) Detail A (showing a y-scale ranging from -610 to -550µm and an x-scale ranging from +140 to +230µm). (c) Detail B (showing a y-scale ranging from +340 to +380µm and an x-scale ranging from -150 to -100µm).

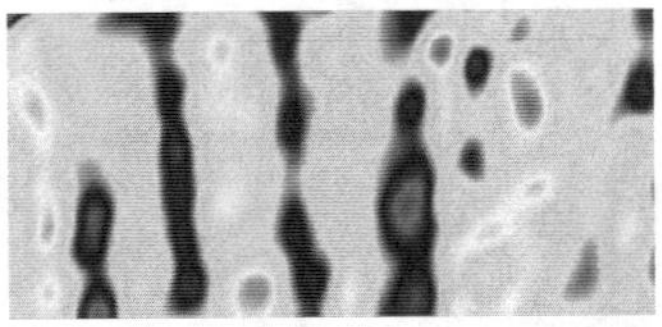

Param.	Primary	Roughness
Sa	0.121μm	0.112μm
Sq	0.149μm	0.136μm
Sz	0.941μm	0.978μm

(a) (b) (c)

Figure 8. (a) 2D-view of the drill tip with the marked area used for the roughness measurement (Dimension of the marked area: 40 x 75μm). (b) Filtered surface used for calculation of roughness parameters. The height values range from -0.5 (dark) to 0.5μm (bright). (c) Roughness Measurement results of (b).

The standard deviations calculated above only provide parts of the measurement uncertainty including the repeatability of the measurement and the uncertainty of the sample. In order to get additional information on the measurement uncertainty for typical tool parameter measurements, the pitch diameter of a calibrated thread plugs has been measured. Our measurement result for a thread plug with the calibrated value of 1.7347mm was 1.7353mm leading to a difference of 600nm. Our measurement result for a thread plug with a diameter of 7.4054mm was 7.4071mm leading to a difference of 1.7μm. The measurement uncertainty of the device used for calibration was +-2.5μm, thus both of our measured values were in the uncertainty range of the calibrated device.

One big advantage of the technology Focus-Variation is the ability of measuring steep flanks. This enables cutting edge measurements with angles up to 20°. For cutting edge measurements it is especially important to measure with a cross section normal to the cutting edge. The cross section of Detail A is shown in Figure 7a. Detail A and B are shown in Figure 7b and 7c. The cutting edge angles are 56.22° for Detail A and 55.83° for Detail B.

In addition to form measurements also roughness measurements can be performed due to the high measurement point density. The measurement shown in Figure 8 is a roughness measurement on the tip of the drill from section 3.1. Figure 8b shows the roughness filtered dataset of the area marked in Figure 8a. Different area based roughness parameters [6] are shown in Figure 8c.

3.2. Measurement of a Micro Milling Cutter

Up to now a big challenge for 360° measurements were components smaller than 100μm. But especially in this range dimensional accuracy is more important than ever. Only a small deviation may result in a short tool life time or in inaccurate manufacturing results. The presented measurement device in combination with the rotation unit enables measurements of components with a diameter as small as 70μm. Shown are two example measurements. Figure 9 and 10 depicts the 360° measurement of the micro milling cutter and shows the result of the cutting tip measurement. The longitudinal cross section is aligned along the rotation axis of the milling cutter.

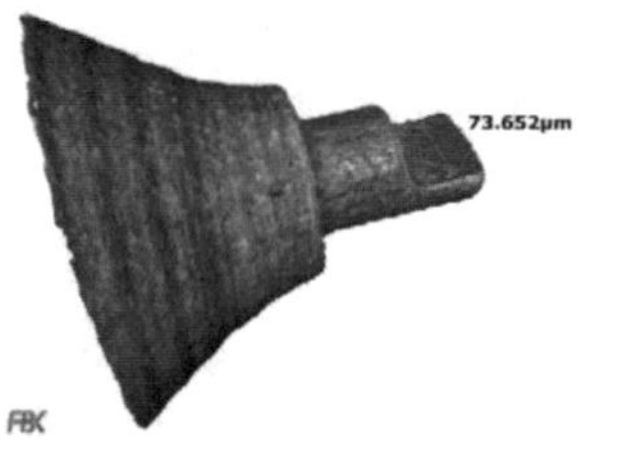

Figure 9. Measured cutting tip diameter by using the contour measurement.

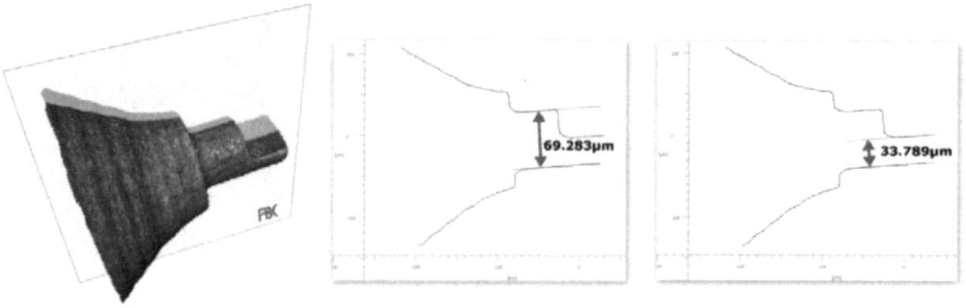

Figure 10. Longitudinal cross section and according measurements performed with the contour measurement.

4. Conclusions

We have demonstrated the measurement of small tools using an optical measurement device based on Focus-Variation. Measurements of the major radius on various positions of a drill with a standard deviation as small as 190nm show the accuracy of the proposed method. Comparisons of pitch diameter measurements of a thread plug to calibrated reference values showed differences that were within the calibration measurement uncertainty of 2.5µm. Measurements on a micro milling cutter demonstrate the ability to measure even tools with diameters below 100µm. In addition to form measurements also roughness measurements (e.g. at the tip of a drill) are applicable due to the high measurement point density of Focus-Variation.

Acknowledgements

We acknowledge the help of Prof. Aurich and DI Guido Schüler for providing micro milling cutters shown in Section 3.2 and supporting information on these tools.

References

[1] Walter Maschinenbau GmbH Tübingen *http://www.walter-machines.com/*

[2] Vermeulen M, Rosielle P C J N and Schellekens P H J 1998 Design of a High-Precision 3D-Coordinate Measuring Machine *Ann. CIRP* **47** Issue 1 447-450

[3] ISO/TS 25178-6 2010 *Geometrical product specifications (GPS) -- Surface texture: Areal -- Part 6: Classification of methods for measuring surface texture*

[4] Leach R K 2009 *Fundamental Principles of Engineering Nanometrology - Micro and Nano Technologies* (Oxford: William Andrew) pp. 142-144

[5] Emuge-Franken (Herausgeber) 2004 *Handbuch der Gewindetechnik und Frästechnik* (Publicis Corporate Publishing)

[6] ISO/TS 25178-2 *Geometrical Product Specifications (GPS) -- Surface texture: Areal -- Part 2: Terms, definitions and surface texture parameters* Draft

A quantitative method to estimate high gloss polished tool steel surfaces

S. Rebeggiani*[1]; B.-G. Rosén[1]; A. Sandberg[2]

[1]Halmstad University, The Functional Surfaces Research Group, Box 823, SE-301 18 HALMSTAD, Sweden
[2]Uddeholms AB, SE-683 85 Hagfors, Sweden

Abstract

Visual estimations are today the most common way to assess the surface quality of moulds and dies; a method that are both subjective and, with today's high demands on surfaces, hardly usable to distinguish between the finest surface qualities. Instead a method based on non-contact 3D-surface texture analysis is suggested. Several types of tool steel samples, manually as well as machine polished, were analysed to study different types of surface defects such as pitting, orange peel and outwardly features. The classification of the defect structures serves as a catalogue where known defects are described. Suggestions of different levels of 'high surface quality' defined in numerical values adapted to high gloss polished tool steel surfaces are presented. The final goal is to develop a new manual that can work as a 'standard' for estimations of tool steel surfaces for steel producers, mould makers, polishers etc.

Keywords: Tool steel, Replication, Surface analysis, Surface characterisation

1. Introduction

Quality controls and/or specifications of the surface condition of injection moulding tools are commonly based on qualitative estimations where reference surfaces are compared to actual ones. Most often even simple roughness parameters like the Ra-value (the arithmetic mean deviation of the assessed profile [1]) are used to better describe the surface condition. However, these kinds of procedures are subjective, and 2D measurements often lack in their ability to describe surface structures. More objective (and standardised) methods for surface evaluations of 'mirror-like' tool steel surfaces are, to the authors knowledge, not existing, therefore new methods/procedures are suggested; well-known surface criteria of tool steel surfaces, collected via visual estimations, were translated into quantitative parameters based on 3D surface measurements.

The first idea is based on the concept to divide the geometrical features on the surface into three components; form, waviness and roughness (see Fig. 1), depending on their spectrum of wavelength. This separation make it possible to distinguish between different types of defect structures, thus a detailed surface characterisation can be performed where type, dimension and distribution of defects can be detected.

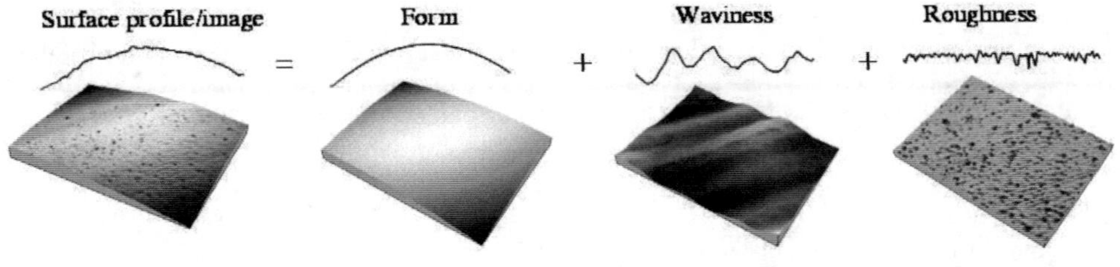

Figure 1: A surface profile divided into its form, waviness and roughness; notice that the profiles/images have various height scales [2].

The second idea is to build up the surface topography by hills, dales and saddle points (see Fig. 2), which is more comparable to visual estimations [3]. A way to visualise useful information is the change tree, which shows the relationships between hills, dales and saddle points; to avoid irrelevant data, Wolf pruning is applied which reduce the amount of hills/dales by merging small contiguous hills/dales into larger ones. In this case, the output values are 'accepted' or 'not accepted' surface quality where the limit values are based on known surface structures and visual estimations of the test samples.

* Sabina Rebeggiani. Tel.: +46 (0)35 167843
E-mail address: sabina.rebeggiani@hh.se (Sabina Rebeggiani)

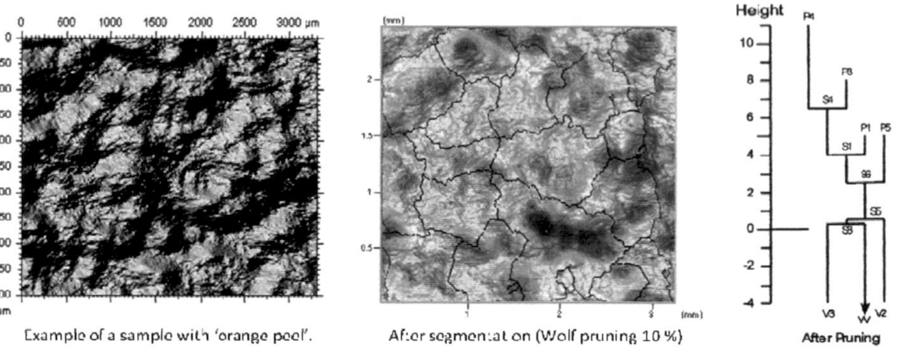

Figure 2: Left; high gloss polished sample with 'orange peel'. Mid; dale segmentation by wolf pruning (10%). Right; example of a change tree [3].

2. Experimental work

Tool steel samples produced via different process routes (powder metallurgy - PM, ingot casting - IC, continuous casting - CC and electro slag remelting - ESR), heat treated to various hardness levels (48 to 60 HRC), were included in the study. The samples, all plane surfaces, were either manual or machine polished in order to achieve a high gloss surface appearance, i.e. a 'mirror-like' surface. A typical surface preparation sequence include one grinding step to secure a plane surface, one to six grinding steps (e.g. with aluminum oxide stones), and finally three to five polishing steps with various diamond abrasives.

2.1 Evaluation technique

A Phase Shift Technology MicroXam[1] optical interferometer operating in the phase-shifting mode (quoted vertical resolution: 0.1 nm; sampling: 1 µm) was used to measure and examine the surfaces at different magnifications. Surface analysis based on the 10x-images (measurements with the magnification of 100) were made in MountainsMap® Premium[2] and SurfStand v5.0[3]; all images were levelled with respect to the least squares plane and a form removing step (polynomial fitting of order 2) was applied to get rid of any remaining curvatures. Surface parameters included in this paper are [1] [3] [4]:

Sq - Root-mean-square deviation of the surface [nm]
Sk - Kernel roughness depth (roughness depth of the core) [nm]
Str - Texture aspect ratio (values near 1 means isotropic surface, values is near 0 means anisotropic surface)
Sdd - Number of dales per unit area [$1/mm^2$]
S5v - Five point pit height [nm]
Mean height of islands (MHoI) - Sum of the maximal heights of the islands above a threshold (which in this case is defined as 2xSk in height from the mean plane) divided by the number of islands [nm]
Mean surface of islands (MSoI) - Sum of the surfaces of the islands above the threshold (which in this case is defined as 2xSk in height from the mean plane) divided by the number of islands [μm^2]

Numerical value	Qualitative evaluation scale	Quality level
1	Surface in very good condition without pitting	Good surface quality
2	Surface in very good condition with min. pitting	Good surface quality
3	Surface in good condition with min. pitting	Good surface quality
4	Surface with min. structure with min. pitting	Accepted surface quality
5	Surface in good condition with some pitting	Accepted surface quality
6	Surface in good condition many pitting	Accepted surface quality
7	Surface with structure and many pitting	Non-accepted surface quality
8	Surface with structure and many deep pitting	Non-accepted surface quality
9	Surface with deep structure and pitting all over	Non-accepted surface quality

Table 1: Summary of the ranking-scale used for the visual estimation (no instruments were used); low values correspond to 'good surface qualities', high values to worse ones. Adapted from [SR lic].

[1]KLA-Tencor, USA, www.kla-tencor.com
[2]Digital Surf, France, www.digitalsurf.com
[3]Huddersfield University/Taylor Hobson Ltd,UK, www.hud.ac.uk/www.taylor-hobson.com

All parameter values given in the tables are average values based on five, randomly chosen measurement points from each sample (except the parameter values given in figure 4, 5 and 6 which are based on one single value - the actual measurement point). A similar study reported that 5-10 measurements were needed to get stable mean values for most parameters (±20% at the 95% confidence level) [3]. The visual estimation, used as a reference to the measurements, was based on two criteria, the condition of the surface structure (e.g. relief and orange peel) and pitting. Part of the samples was estimated in a well-known polishing shop, part of them at Halmstad University. To be able to compare this qualitative evaluation to the calculated surface parameters, the results were translated into numerical values (see table 1).

3. Results and discussion

3.1 Defect classification – defining the geometry of a defect

Based on interviews, questionnaires [5], literature studies (e.g. [6-9]) and analyses of test samples a summary of defect structures was presented in [2]. That work has further been developed into a more complete defect classification in this study (see Fig. 3). All defect structures are presented as shown in the example for the inwardly directed defect pitting (Fig. 3, right). This particular defect is a well-known problem within industry and, as it is often spread over the majority of the surface, it is relatively easy to detect by a trained (naked) eye. One way to separate pitting from other inwardly defects is to calculate the MHoI, describing the mean height of the defects (note: the images are inverted), and MSoI, describing the mean defect area. As can be seen in figure 4 the pitting are smaller than the holes (have lower MSoI values), however they are not always more shallow as in the example in figure 3.

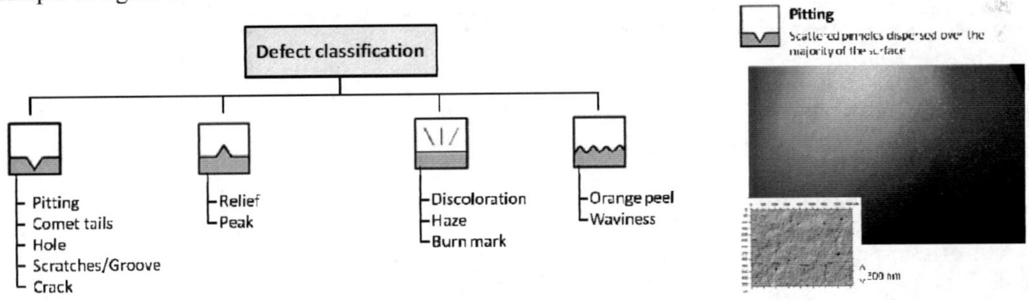

Figure 3: Left; a schematic view of included defect types, from the left: inwardly directed defects, outwardly directed defects, areas that appear different compared to its surrounding, and wavy textures. Right; an example of how 'pitting' can be described in a defect chart.

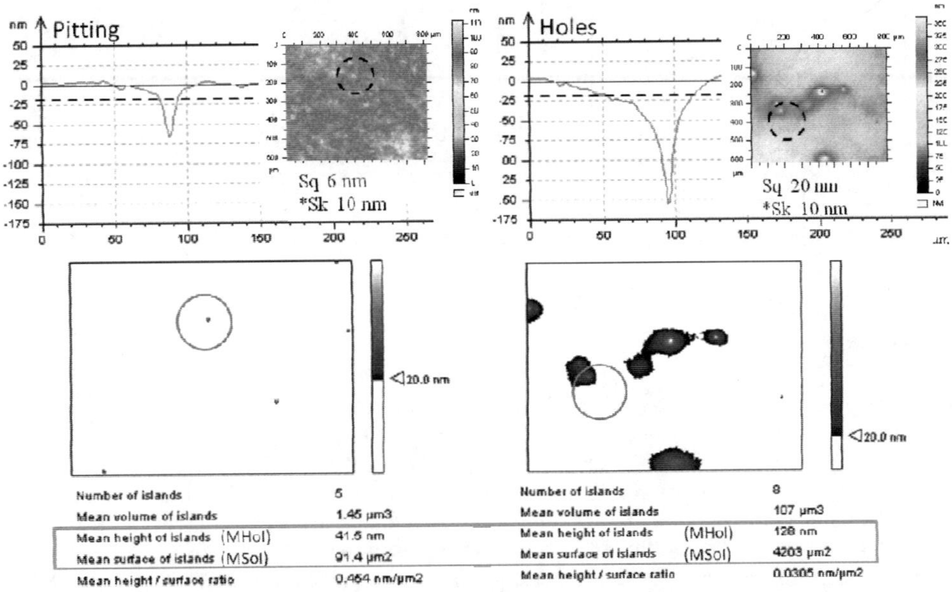

Figure 4: An example of how pitting can be separated from holes by the MSoI value (lower images inverted). *Given Sk values are based on filtered surfaces; Robust Gaussian, cut-off 150 µm.

3.2 Quantitative method for extraction of defects

This stepwise analysis method is described in figure 5, where a non-accepted sample is serving as an example; after the levelling and form removal operation, the image is further modified to separate any wavy textures from out- and inwardly directed defects. This enables a detailed study of the defect structures included in the defect chart with one exception - the 'areas that appear different compared to its surrounding', which need to be further studied.

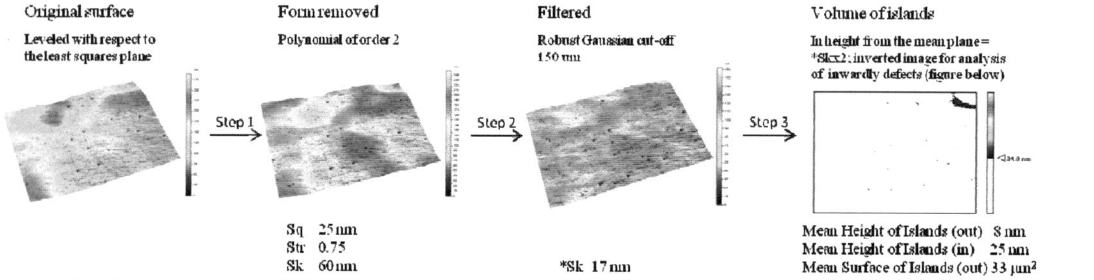

Acceptance levels /parameter	Sq [nm]	Str	Sk [nm] (step 1)	MHoI out [nm]	MHoI in [nm]	MSoI out [μm²]
Good surface quality	< 12	≥ 0,4	< 25	< 5	< 5	< 30
Accepted surface quality	12-20	≥ 0.4	25-35	5-15	5-15	30-200
Non-accepted surface quality	> 20	< 0.4	> 35	> 15	> 15	> 200
Type of defect structure:	to get an overview of the surface roughness	to detect anisotropic structures (e.g. Scratches)	to detect orange peel	to detect outwardly features	to detect inwardly features	to detect relief structures

Figure 5: Based on one single interferometer measurement, a sample ranked as 7 (see table 1) in the visual estimation, due to orange peel and pitting in the surface, was analysed in three steps. The table shows suggested acceptance levels; grey cells correspond to the values of the example surface – i.e. it fails due to orange peel and inwardly defects.

Figure 5 also presents numerical values for three acceptance levels of high gloss polished tool steel surfaces in order to define an objective estimation of the surfaces. The values are chosen with respect to the visual estimations of the samples, i.e. they are based on the opinions of the polisher. Table 2 summarises results based on included samples, which cover both accepted and non-accepted surface qualities. As can be seen, different numerical values correspond to different types of defect structures; coloured cells point out interesting values.

Sample (process route)	Visual estimation	Sq [nm]	Str	Sk [nm] (step 1)	MHoI out [nm]	MSoI out [μm²]	MHoI in [nm]	MSoI in [μm²]	S5v/Sdd	Comments
KSS1 (ESR)	2	4	0,6	10	8	20	7	9	0,06	Some few peaks & pin-holes detected
KVN (PM)	1	7	0,6	19	3	17	0	0	0,06	Very good surface quality
KIHH (IC)	5	15	0,3	20	21	106	77	409	*	Holes detected
S58 (IC)	9	6	0,6	11	5	417	9	15	1,8	Relief structure
S89 (CC)	8	1	0,5	10	13	109	17	62	1,3	Pitting all over
K2C (ESR)	7	22	0,7	57	11	1550	8	29	1,1	Orange peel
K2D (ESR)	8	11	0,1	27	1	1	12	35	0,2	Ok surface condition and no pitting, but non-glossy surface

Table 2: Summary of analysed samples; numerical values, acceptance levels and defects are presented sample by sample.

3.3 Quantitative surface estimation

To get a more general evaluation of the surface, i.e. to limit the output to one single value - accepted or non-accepted – the procedure in figure 6 can be used. The surface is modified in the same way as before, but instead of using a Robust Gaussian filter, Wolf pruning is used which in this case marks the borders between dales. Defect free (or nearly defect free) surfaces consist of many small dales, surfaces with defects of less and bigger ones (provided a Wolf pruning of 10%). To separate accepted and non-accepted surfaces the S5v value (≈ valley depth) is divided by the Sdd value (density of dales) – a value below 1 means an accepted surface quality. This is because deeper inwardly or outwardly directed defects affect the segmentation, and so lower the Sdd value and increase the S5v value; the quotient will exceed 1. A summary of the results are presented in table 2 above; the method fails in some few cases where no values are presented - it needs to be further improved.

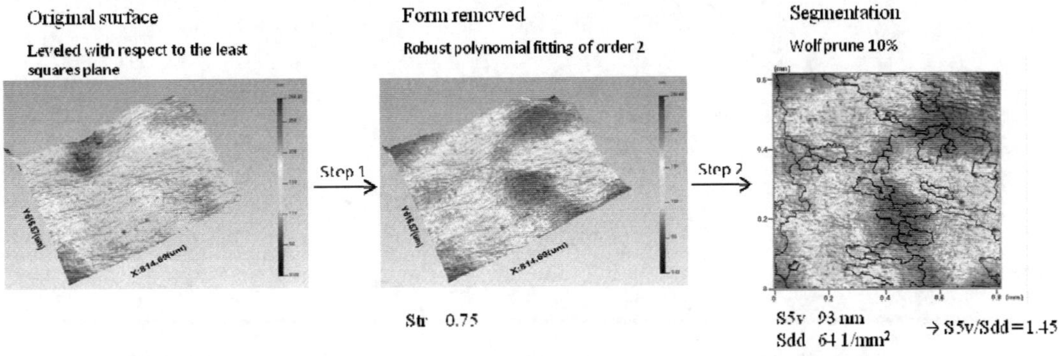

Acceptance levels / parameter	Str	S5v/Sdd
Good/accepted surface quality	> 0.4	< 1
Non-accepted surface quality	< 0.4	≥ 1
Comments	To detect anisotropic structures	Based on dales

Figure 6: Based on one single interferometer measurement, a sample ranked as 7 (see table 1) in the visual estimation, due to orange peel and pitting in the surface, was analysed in two steps. Suggested acceptance levels are presented; grey cells correspond to the values of the example surface – i.e. it fails due to too extensive defect structures.

4. Conclusions

Even though further analyses are needed it can be concluded that the methods presented can be used as objective surface estimations; non-accepted defect structures were detected with help of interferometer measurements and defined by a few surface parameters. E.g. pitting, which is an inwardly directed defect, could be defined by calculating the MSoI- and MHoI values.

5. Future work

Reference samples and moulds, with defined defect structures, will be studied in order to measure in what extend the defects are transferred into different types of plastic plaques; the goal is to find the size/shape limit where the defects do not affect the final quality of the plastic surfaces. Further studies will also include the 'areas that appear different compared to its surrounding', and improvements of the suggested methods.

Acknowledgements

The authors would like to thank: Uddeholms AB for their financial support and cooperation; Prof. Liam Blunt and his colleges at Huddersfield University for providing me their software SurfStand; Digital Surf for providing us the software MountainsMap Premium; Kiefer GmbH for preparation of the samples; all colleagues in The Functional Surfaces Research Group at Halmstad University.

References

[1] Software: MountainsMap Premium, Digital Surf, France, www.digitalsurf.com
[2] S. Rebeggiani, Polishability of Tool Steels – Characterisation of High Gloss Polished Tool Steels. Licentiate dissertation, Chalmers University of Technology, Gothenburg, 2009.
[3] L. Blunt, X. Jiang. Advanced Techniques for Assessment Surface Topography – Development of a Basis for 3D Surface Texture Standards "SURFSTAND", Kogan Page Science, London and Sterling, England, 2003.
[4] Software: SurfStand v.5.0, Huddersfield University/Taylor Hobson Ltd, UK/UK, www.hud.ac.uk/www.taylor-hobson.com
[5] S. Rebeggiani, Polishability of tool steels, Forskningsmeddelande FM09-100-1, 2009.
[6] G. Kölzer, D. Lucht. Das JOKE Buch, Joisten and Kettenbaum GmbH & Co. KG, Bergisch Gladbach, Germany, 2003.
[7] A. Botzian, K. Argmann. Modern methods for mirror polishing, Botzian & Kirch GmbH, Bergisch Gladbach, Germany, 2009.
[8] www.struers.com (101215)
[9] ISO 8785:1998, In: Handbook 539:1999 – Ytstruktur (Utgåva 1), Svensk Material- & Mekanstandard, ISBN 91-7162-478-3, 1999.

Journal of Physics: Conference Series **311** (2011) 012005 doi:10.1088/1742-6596/311/1/012005

Calibration of the geometrical characteristics of areal surface topography measuring instruments

C L Giusca[1, 2], R K Leach[1], F Helery[3], T Gutauskas[4]

[1]National Physical Laboratory, Hampton Road, Teddington TW11 0LW, UK

[2]School of Computing & Engineering, University of Huddersfield, UK

[3]Ecole Nationale Supérieure d'Arts et Métiers, France

[3]Imperial College London, UK

claudiu.giusca@npl.co.uk

Abstract. The use of areal surface topography measuring instruments has increased significantly over the past ten years as industry starts to embrace the use of surface structuring to affect the function of a component. This has led to a range of areal surface topography measuring instruments being developed and becoming available commercially. For such instruments to be used as part of quality control during production, it is essential for them to be calibrated according to international standards. The ISO 25178 suite of specification standards on areal surface topography measurement presents a series of tests that can be used to calibrate the metrological characteristics of an areal surface topography measuring instrument. Calibration artefacts and test procedures have been developed that are compliant with ISO 25178. The material measures include crossed gratings, resolution artefacts and pseudo-random surfaces. Traceability is achieved through the NPL Areal Instrument – a primary stylus-based instrument that uses laser interferometers to measure the displacement of the stylus tip. Good practice guides on areal calibration have also been drafted for stylus instruments, coherence scanning interferometers, scanning confocal microscopes and focus variation instruments.

1. Introduction

An areal surface topography measuring instrument provides a three dimensional (3D) map of a surface. The 3D map is made up of a set of points measured with respect to three orthogonal length scales. The scales of the areal surface topography measuring instrument should be aligned to the axes of a Cartesian co-ordinate system. The axes are physically realized by various components that are part of the metrological loop of the instrument. Hence the quality and the mutual position of these components partially confer the quality of the co-ordinate measurements. The co-ordinate measurements produced by areal surface topography measuring instruments are also affected by other influence factors such as ambient temperature, mechanical noise and electrical noise. The effect of a single influence factor, or a combination of influence factors, on the quality of the areal measurements should be quantified by experimentally determining the metrological characteristics of the instrument. Amongst these characteristics of areal surface topography measuring instruments are the noise of the

Journal of Physics: Conference Series **311** (2011) 012005 doi:10.1088/1742-6596/311/1/012005

instrument; the linearity, amplification and resolution of the scales; the deviation from flatness of the areal reference and the perpendicularity of the axes. The effect of the influence factors on an areal measurement can have a different magnitude for different sizes of measuring area and sampling distance, that is to say the measurement bandwidth. The choice of measurement bandwidth is application dependent and is based on the selection of S-filters and L-filters/F-operators, each having a range of preset values called nesting indexes [1, 2]. The calibration of the instrument should be performed using the same conditions as those used on a daily basis.

This paper presents a way of calibrating the scales of areal surface topography measuring instruments. The calibration of the scales consists of a series of relatively simple tasks that evaluate the magnitude of the uncertainty associated with the metrological characteristics of the instruments whilst assuming well-defined measuring conditions. The calibration process also requires the use of artefacts designed for calibrating surface topography measuring instruments. A coherence scanning interferometer (CSI) has been used as an example instrument [2, 3]. The CSI was equipped with a one mega-pixel camera, a Mirau type 50× magnification objective lens with a working field of view of 0.35 mm by 0.35 mm and a 0.1 mm range piezo-electric scanner. The areal data was analyzed using commercially available surface texture analysis software (TalyMap version 5).

2. Calibration steps

The main input quantities that influence the uncertainty associated with the co-ordinate measurements produced by areal surface topography measuring instruments are: measurement noise; flatness; amplification coefficient and the linearity of the scales (z and lateral scales); orthogonality of the axes; resolution.

2.1. Measurement noise

The magnitude of the measurement noise can be estimated by measuring the root mean square (RMS) of the scale limited surface Sq on a flat surface of less than or equal to 30 nm peak to valley (VDI/VDE 2617 [4]). The difficulty of the test is in separating the effect of the roughness of the flat surface from the instrument noise. VDI/VDE 2617 presents a method of separating the noise from the flat surface roughness. The method consists of subtracting two repeated measurements from each other. Another technique of error separation was developed elsewhere [5]. This technique is based on the assumption that the instrument noise contribution to the RMS of the average surface obtained from multiple measurements performed at the same location on the flat will decrease by the square root of the number of repeat measurements, so that the noise of the instrument can be estimated by equation (1)

$$Sq_{noise} = \sqrt{\frac{Sq^2 - Sq_n^2}{1 - \frac{1}{n}}} \tag{1}$$

where Sq is the measured RMS after one measurement, n is the number of repeated measurements, Sq_n is the measured RMS of the averaged surface and Sq_{noise} is the noise of the instrument.

The latter method was chosen to measure the noise of the CSI instrument because the instrument's acquisition software allowed for repeated measurements and averaging. The noise results of a transparent glass flat measured using the CSI showed that two repeated measurements ($n = 2$) are enough to determine the noise of the instrument. The Sq_{noise} does not differ significantly (less than 0.1 %) with the increase in the number of repeated measurements.

2.2. Flatness

An important quality of any areal surface topography measuring instrument is the flatness of its areal reference. Similarly to the measurement noise test, the flatness test is performed on a flat surface but the parameter that quantifies the magnitude of the flatness effect is the maximum height of the scale limited surface (Sz). Unlike Sq, the value of Sz is highly sensitive to local height variations such as scratches or dirt. It is, therefore, difficult to completely separate the contribution of the instrument from that of the flat and other spurious measurement data. One way to overcome these issues is to measure the topography of the flat at different locations [4] without changing the instrument setup and to average the height measurement of each point of the topography. The contribution of the flat and any spurious data should diminish whereas the quality of the areal reference should be preserved. It is difficult to recommend the exact number of repeated measurements because the number depends on the rate at which the value of Sz stabilises. The measurements can be repeated until the value of Sz stabilizes.

Sz is 3.26 nm for the CSI instrument after averaging ten measurements. In this case the magnitude of Sz is dominated by a virtual scratch present in the areal reference of the instrument (see figure 1), which is inherited from the flat that was used to adjust the instrument. A better quality flat would have produced a better quality areal reference. Sz is 4.69 nm after averaging four measurements, that is sufficient if the measurement uncertainty along the z direction does not need to be better than 10 nm.

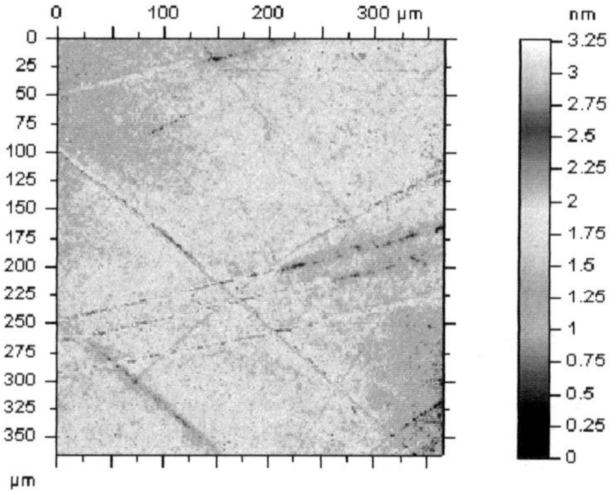

Figure 1 Flatness of the areal reference of the CSI used to measure a transparent glass flat - result after ten repeated and averaged measurements.

2.3. z axis calibration

The z axis calibration consists of a series of measurements of different step height artefacts with various depths to establish the relationship between the ideal response curve and the instrument response curve. This calibration provides information about the z axis linearity and amplification coefficient. The linearity of the z axis is given by the maximum deviation of the instrument response curve from the linear curve where the slope is the amplification coefficient.

The step height artefacts should cover the entire z axis range of the instrument or at least they should range from the minimum to the maximum height of interest. The linearity and amplification coefficient can be extracted from the measurement results of multiple calibrated step heights of different values by simply performing a linear fit.

Often the z axis calibration is performed using a single step artefact that is also used to adjust the software of the instrument. This situation is potentially very dangerous because it could only shift the response curve in such a way that it crosses the ideal response curve only at that particular measured point. Unless the response curve is perfectly linear, one calibration artefact does not provide sufficient information about the quality of the instrument's z axis scale and it will underestimate its contribution into the uncertainty calculations. The z axis scale should be calibrated at regular intervals because the instrument characteristics could change over time.

The squareness between the areal reference and the z axis behaves as an amplification error.

2.4. *xy* scales calibration

Traditionally the calibration of the lateral axis of a roughness measuring instrument is performed using a grating with a known pitch. The pitch measuring technique can be applied to calibrate instruments that measure areal topography but the analysis has to be performed in profile mode. The drawback of pitch measurements is that they only estimate the local characteristic of the instrument's scale and do not give information about the instrument response curve. Areal material measures such as cross gratings [6] or pyramidal structures [7] are better suited for calibration of the lateral scales of areal topography measuring instruments.

The x and y axes amplification and the squareness can be measured using a calibrated cross grating artefact. By measuring the positions of the centre of gravity of the cross grating's squares with a traceable areal surface topography measuring instrument [8, 9] allows for stable and traceable length measurements along the x and y axes. Additionally, the squareness of the x and y axes can be measured by measuring the angle between two nominally orthogonal rows of square holes whose squareness is known. The orientation of each row of squares can be calculated by fitting a line though the centre of gravity of the corresponding squares.

2.5. Resolution

The resolution of the lateral scales is defined as the smallest lateral separation between two points that can be distinguished. This is a useful definition of the lateral resolution for a 2D microscope or when making lateral measurements. However, for areal surface topography measurements, where the distance between two adjacent points could affect their relative height difference, this definition becomes impracticable. The width limit for full height transmission has been defined [10] to overcome the problem with the lateral resolution definition but is still under debate. Experimentally the width limit for full height transmission is measured on gratings or crossed gratings with the pitch value close to the resolution of the instrument. A 3D star pattern [6, 11] can be used to find the approximate value of the width limit for full height transmission before measuring a grating with a predefined pitch.

Other definitions of lateral resolution could be used where the width limit for full height transmission is not appropriate for a specific application. ISO/TC213-WG16 is looking to introduce an umbrella term *spatial height resolution* defined as the ability of a surface topography measuring instrument to distinguish closely spaced surface features [3].

3. Uncertainty

The calculation of uncertainties for surface topography measuring instruments is a very complex task that is often only carried out at National Measurement Institutes. It is often possible to calculate the instrument uncertainty, that is to say the uncertainty in measuring (x, y, z), but when the effect of the surface is taken into account this uncertainty value may significantly increase, often in an unpredictable manner. Where possible the guidelines in the GUM [12] should be applied to calculate instrument uncertainties and the effect of the surface should be considered in as pragmatic a manner as possible. Until now little work has been carried out for the uncertainty associated with areal parameters [13].

When the instrument uncertainty has been calculated it is then often necessary to find the uncertainty in a parameter calculation. Once again this is far from trivial and often the guidelines in the GUM cannot be easily applied [14]. The problem is that for roughness parameters, some characteristics of a roughness measuring instrument have an obvious influence on a roughness parameter, but for others this is highly unclear [15].

4. Conclusions

This paper presents a simple way of calibrating the scales of an areal surface topography measuring instrument. The calibration consists of measuring the noise and flatness of the instrument, amplification coefficients, linearity and squareness of the scales. More details on calibration procedures and calibration results will be published in several NPL Good Practice Guides.

It is important to point out that measuring the above metrological characteristics is not enough to qualify an areal instrument for the measurement of complex, rough surfaces. For this, the spatial frequency response of the instrument needs to be calibrated and such methods are still in their infancy [16, 17, 18]. The effect of the filters employed for a specific application also has to be studied. For example the CSI instrument used in this paper can accurately measure very small step heights because the mathematical algorithm used to estimate the step value reduces significantly the effect of the noise and flatness errors.

Acknowledgements

The project was funded by the BIS National Measurement System Engineering Measurement Programme 2008 to 2011.

References

[1] ISO/DIS 25178-3:2010 Geometrical product specifications (GPS) - Surface texture: Areal - Part 3: Specification operators, International Organization for Standardization

[2] Leach R K 2001 *Fundamental principles of engineering nanometrology* (first ed., Amsterdam Elsevier)

[3] ISO/DIS 25178-603:2010 Geometrical product specifications (GPS) - Surface texture: Areal - Part 603: Nominal characteristics of non-contact (phase shifting interferometric microscopy) instruments, International Organization for Standardization.

[4] VDI/VDE 2617 Part 6.2 2004 Accuracy of coordinate measuring machines. Characteristics and their testing. Guideline for the application of DIN EN ISO 10360 to coordinate measuring machines with optical distance sensors

[5] Haitjema H and Morel M A A 2005 Noise bias removal in profile measurements *Measurement* **38** 21-29

[6] Leach R K, Chetwynd D, Blunt L, Haycocks J, Harris O, Jackson K, Oldfied S and Reilly R 2006 Recent advances in traceable nanoscale dimension and force metrology in the UK *Meas. Sci. Technol.* **17** 467-476

[7] Ritter M, Dziomba T, Kranzmann A and Koenders L 2007 A landmark-based 3D calibration strategy for SPM *Meas. Sci. Technol.* **18** 404-414

[8] Leach R K, Giusca C L and Naoi K 2009 Development and characterisation of a new instrument for the traceable measurement of areal surface texture *Meas. Sci. Technol.* **20** 125102

[9] Thomsen-Schmidt P and Krüger-Sehm R 2008 Calibration of an electromagnetic force compensation system *Proc. XII Int. Colloq. Surf (Chemnitz, Germany, 28-29 Jan.)* p 323-327

[10] ISO 25178-601:2010 Geometrical product specifications (GPS) - Surface texture: Areal - Part 601: Nominal characteristics of contact (stylus) instruments, International Organization for Standardization

[11] Weckenmann A,Tan Ö, Hoffmann J and Sun Z 2009 Practice-oriented evaluation of lateral resolution for micro- and nanometre measurement techniques *Meas. Sci. Technol.* **20** 065103

[12] BIPM, IEC, IFCC, ISO, IUPAC, IUPAP and OIML 2008 Guide to the Expression of Uncertainty in Measurement, Bureau International des Poids et Mesures JCGM 100

[13] Morel M A A and Haitjema H 2001 Calculation of 3D roughness measurement uncertainty with virtual surfaces, Proc. IMEKO, Cairo, Egypt 1-5.

[14] Harris P M, Leach R K and Giusca C L 2010 Uncertainty evaluation for the calculation of a surface texture parameter in the profile case *NPL Technical Report MS 8*

[15] Haitjema H 1998 Uncertainty analysis of roughness standard calibration using stylus

instruments *Precision Engineering* **22** 110-119
[16] Yahschuk V V, McKinney W R and Takacs P Z 2008 Binary pseudorandom grating standard for calibration of surface profilometers *Opt. Eng.* **47** 073602
[17] Polodhi K, Coupland J A and Leach R K 2010 A linear model of fringe generation and analysis in coherence scanning interferometry *Proc. ASPE Summer Topical Meeting on Precision Interferometric Metrolog, (Ashville, USA)*
[18] Leach R K and Haitjema H 2010 Bandwidth characteristics and comparisons of surface texture measuring instruments *Meas. Sci. Technol.* **21** 032001

Characterization of nano-structured surfaces by EUV scatterometry

F Scholze[1], A Kato, C Laubis

Physikalisch-Technische Bundesanstalt, Abbestrasse 2-12, D-10587 Berlin, Germany

E-mail: frank.scholze@ptb.de

Abstract Scatterometry in the UV and the VIS is widely used for the inspection of nano-structured surfaces in, e.g. quantitative wafer metrology in the semiconductor industry. As the structures become progressively smaller, ever fewer propagating diffraction orders exist. Consequently, these few orders do not carry enough information about the structure any more. Particularly the reconstruction of structures without detailed a priori knowledge becomes impossible due to this lack of information. The Physikalisch-Technische Bundesanstalt (PTB) has developed extreme UV (EUV) scatterometry using its EUV reflectometry facility at the electron storage ring BESSY II. The short wavelength of EUV at around 13 nm is advantageous since it provides more propagating diffraction orders compared to the longer wavelength radiation. The short wavelength also increases the sensitivity to small structural features, particularly roughness. We present the application of EUV scatterometry for the characterization of absorber lines with a trapezoidal cross section on semiconductor photo masks and for the characterization of ultra-smooth surfaces of EUV multilayer mirrors. It is shown that structure roughness significantly impacts the scattered diffraction intensities from structured surfaces and must be included in the structure reconstruction algorithms using inverse modelling by FEM. Estimates for the changes in reconstructed geometrical parameters induced by roughness are presented.

1. Introduction

Scatterometry is an indirect measuring method. It relies on the deduction of the properties to be measured, i.e. the geometry of the scattering object, from the influence of this object on the quantities measured, i.e. the diffracted light. There are two principal

[1] To whom correspondence should be addressed.

approaches. Firstly, the description of diffuse optical scattering from randomly rough surfaces uses disturbance theory and statistical methods to derive expressions for the intensity and directional distribution of the scattered light. The basic relations are outlined in textbooks, e.g. [1]. It should be noted, that the range of spatial frequencies in the surface structure that can be investigated by this method is limited by the wavelength of the radiation, as the existence of real diffraction orders is a precondition for the application of the theory. Secondly, the optical properties of arbitrary structures can be calculated using different methods like FDTD, RCWA, and FEM Maxwell solvers. For reasons of resources and time, however, only rather small domains can be directly computed. Thus this approach is usually applied to periodic structures and a single period of the structure is calculated using periodic boundary conditions to extend it to larger areas. In this sense, scatterometry has become one of the most common techniques in quantitative wafer metrology [2]. Various methods are used, mostly in the spectral range between 200 nm and 1000 nm. As the structures become smaller and smaller, fewer and fewer propagating diffraction orders exist and, consequently, they no longer carry enough information about the structure. For structure pitch measurements by optical goniometry, the smallest pitch that can be determined in backscattering Littrow (incoming and diffracted beam are parallel) geometry is half the wavelength. For wavelengths longer than twice the structure pitch, no diffracted orders exist for any angle of incidence. In this range, the structures work as an effective index medium [3] and detailed a priori knowledge of the geometry is mandatory to derive any structural parameters from the measured effective index. Shorter wavelengths are, therefore, desirable to establish scatterometry as a traceable and absolute metrological method for dimensional measurements of nano-structured surfaces.

At PTB, both approaches are used. Measurements of the diffuse scatter from EUV multilayer mirrors were used to derive the parameters for the interface roughness replication during the layer deposition process [4]. Detailed EUV scatterometric investigations of periodic structures were done at absorber lines of an EUV photomask [5]. It was shown that it is feasible to derive the information on the line profile in periodic areas of lines and spaces by means of rigorous numerical modelling. The comparison with CD measurements using a scanning electron microscope (CD SEM) showed a good correlation of the results [6,7]. Regarding the side wall angle (SWA) of the structures, a remarkable discrepancy of the scatterometry values (below 80°) and the values obtained by AFM measurements (above 85°) was observed. This could recently be explained by the influence of line roughness on the diffraction intensity distribution in angular resolved scatterometry [8]. The latter approach introduced a combination of the above-mentioned complementary interpretations of scatterometry as either a perturbation of specular reflection from a smooth surface or diffraction from a periodically structured surface. Here we comprehensively present recent results for both approaches of scatterometry.

2. Experimental set-up

The data presented here were measured at the EUV reflectometry facility of PTB [9,10] in its laboratory at the storage ring BESSY II. The PTB's soft X-ray radiometry beamline [11] uses a plane grating monochromator which covers the spectral range from 0.7 nm to

35 nm and was particularly designed to provide highly collimated radiation. Therefore, it uses a long focal length of 8 m in the monochromator and the focusing in the non-dispersive direction is provided by the collecting pre-mirror with a focal length of 17 m. We achieve a collimation of the radiation in the experimental station to better than 200 µrad and the scatter halo of the beam can be suppressed to below 10^{-5} relative intensity at an angle of only 1.7 mrad to the central beam. Using a CCD detector with a 13 µm pixel size at a 1.4 m distance from the sample, we achieve a 10 µrad angular resolution in the detection and can cover about 12 orders of magnitude in intensity from the specular peak to the lowest diffuse scatter intensity, if blocking the specular beam. The measurement scheme for angular resolved scatterometry at periodic structures is presented in Figure 1. Usually, the sample is set at a fixed angle to the incoming radiation and the detector is moved. For the measurement of diffuse scatter from randomly rough surfaces, we use a CCD mounted at a fixed angle with respect to the incoming beam and the sample is rocked, see Figure 2.

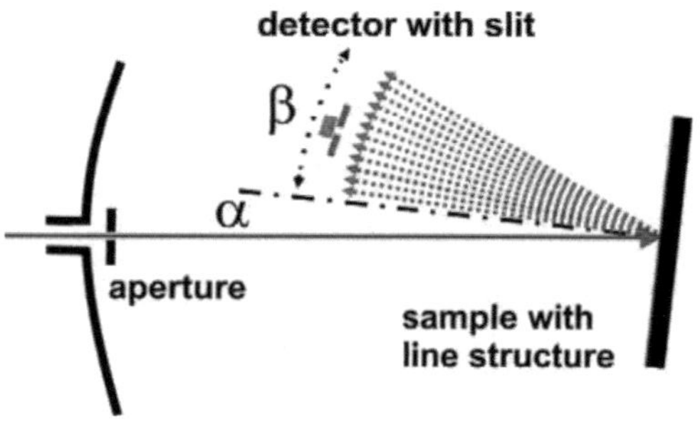

Figure 1. Scheme of scatterometry measurements: The detector angle β is scanned at a fixed incidence angle α.

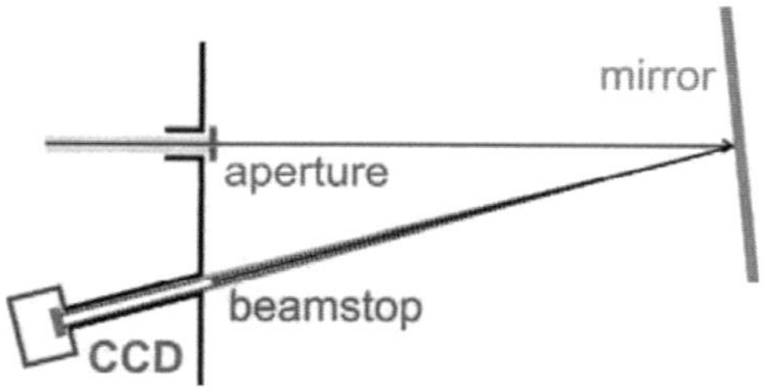

Figure 2. Scheme of small angle diffuse scatter measurement with a CCD detector.

3. Diffuse scatter from super polished mirrors

For short wavelengths below about 30 nm, high reflectance at near normal incidence can only be achieved by multilayer interference reflective coatings. In order to work properly, the interface roughness of these coatings must be extremely low, in the range below 1 nm. For imaging mirrors in EUV optics for semiconductor lithography, roughness amplitudes of well below 0.1 nm (rms) are required over a very broad range of spatial frequencies [12]. These low roughness amplitudes can hardly be directly measured using long wavelength optical radiation scattering. Even more importantly, the resonant reflection from the multilayer coating must be characterized. This requires at-wavelength EUV measurements. In order to access the flare in the optical image of an EUV projection lens, very small scattering angles are most relevant.

As an example, we show data from our reference mirrors used for monitoring the measurement stability of our reflectometer, coated by Fraunhofer IWS in 2001. Figure 3 shows a series of images obtained with the highly collimated beam. The footprint of the

Journal of Physics: Conference Series **311** (2011) 012006 doi:10.1088/1742-6596/311/1/012006

specular reflected beam is extremely narrow and does not have a discernible halo of scatter radiation from the beamline. With a small beamstop blocking only the angular range 2 mrad around the specular direction, the integration time can be increased by a factor of 1000 and the diffuse scatter from the mirror with an almost isotropic distribution is detected. The black shadows in the image arise from the wires used to support the beamstop. A further significant increase in dynamic range by a factor of 20 is obtained by increasing the beamstop diameter to 4 mrad. Note that also the horizontal and vertical scatter lines which arise from beamline apertures are significantly reduced with the larger beamstop. Nevertheless, the scatter from the beamline and from the mirror can be clearly discriminated by the different diameters of the shadow of the beamstop. The origin of the beamline scatter is further from the beamstop than the mirror and thus creates a smaller shadow.

Figure 3. Diffuse scatter of a multilayer coated EUV mirror measured at a 13 nm wavelength with a CCD. The figures are taken with increasing integration time, 0.2 s, 200 s, and 4000 s, from left to right, respectively.
For the pictures with longer integration time, the central specular beam was blocked by a beamstop of 2 mm, or respectively 4 mm, in diameter.

Within the almost isotropic distribution of the diffuse scatter, we resolve a mesh-like fine structure. The very low divergence of the radiation provides this high angular resolution. Quasi-periodic structures in the scatter corresponding to spatial roughness frequencies in the range of 100 μm are resolved with about 100 μrad angular resolution (Figure 4, right). The influence of the collimation of the incident radiation is demonstrated by comparison to the image measured with a larger aperture in the beamline (Figure 4, left). There, the high frequency intensity fluctuations are spread out to a speckle-like pattern. The significance of these small angle intensity variations is also illustrated in Figure 5, where the PSD's derived from the images with 200 s and 4000 s integration time almost completely overlap, indicating that the fluctuations are not signal noise. Images with high angular resolution, as shown in Figure 4, can be used to study geometrical properties of the surface roughness which are not completely random.

Journal of Physics: Conference Series **311** (2011) 012006 doi:10.1088/1742-6596/311/1/012006

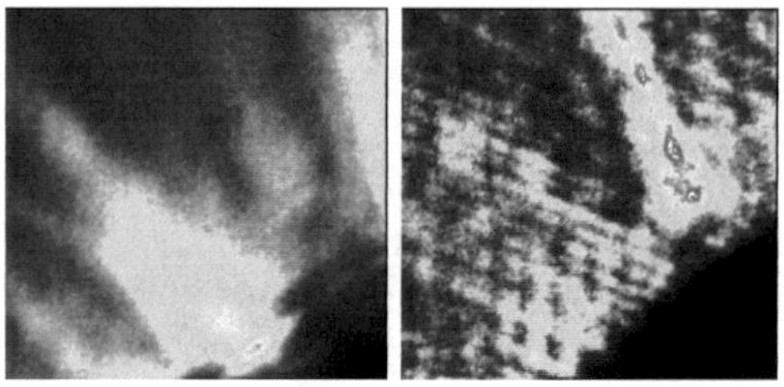

Figure 4. Details of diffuse scatter close to the central beam stop measured with a divergence of 2 mrad horizontally and 0.25 mrad vertically, left, and 0.4 mrad horizontally and 0.15 mrad vertically, right. The angular spread of the lamellar intensity structures observed with low divergence is about 100 μrad.

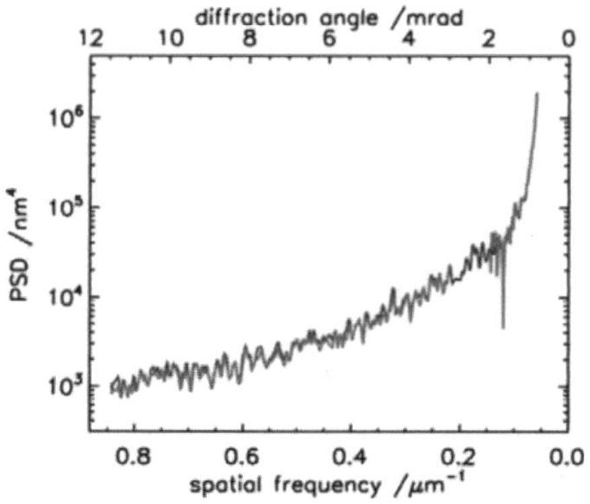

Figure 5. PSD derived from CCD data shown in Figure 3. The spatial frequency range from about 1 μm^{-1} to 0.1 μm^{-1} is accessible with a single CCD image. The colours red, blue, and green correspond to the integration times of 0.2 s, 200 s, and 4000 s, respectively.

4. Angle resolved scatterometry at periodic line structures

The experimental data presented here are obtained from large periodically structured areas at an EUV photomask [5]. It was a negative structure with 540 nm wide and 72 nm thick absorber lines at a pitch of 720 nm, i.e. semi-dense bright lines with 180 nm nominal CD. The measurement method used was angular resolved scatterometry. The scan range of the detector angle β was from -5° to 25°, see Figure 1. The -10th to +14th diffraction order were measured, see Figure 6. As the EUV radiation is resonant with the underlying multilayer reflective coating, the measured diffraction intensities strongly depend on the actual wavelength. We measured at three adjacent wavelengths around the resonance wavelength of 13.65 nm of the multilayer mirror.

The geometrical properties, i.e. the width at the bottom and at the top and the height of the absorber lines were reconstructed by inverse modelling using FEM [6,7]. Three aspects are to be investigated for a proper estimation of the measurement uncertainties. Firstly, the measured diffraction intensities have measurement uncertainties, resulting in uncertainties of about 0.5 nm, or 0.5° SWA for the reconstruction. Secondly, it must be ensured that the model represents the measurement object sufficiently well. A complete description of the sample geometry requires a principally infinite number of parameters, and will, thus, always be an idealization of the real object. Usually it is not possible to simultaneously determine all these parameters from the measured data. Therefore, an a priori judgment is needed as to which parameters are to be kept fixed and which must be adjusted. These considerations contribute uncertainties of typically 1.5 nm, or 1.5° SWA. A detailed investigation of these first two aspects is given in [13]. Thirdly, there are assumptions in the model which are inherent properties of the mathematical methods used. The assumption that the structures are periodic over a sufficiently large area, i.e. the measurement field, is of the most importance. This assumption is needed to restrict the calculation domain to a size which can be treated numerically within a reasonable time. Therefore, any influence of roughness and other irregularities of the structures cannot be estimated straightforwardly by using the same mathematical algorithms. Using the relations for the impact of line width roughness (LWR) and line edge roughness (LER) on the diffraction intensities recently derived [8], we could show that the roughness induced damping of the diffraction intensities (Figure 7) severely impacts the reconstructed SWA while the mean of the top and the bottom CD is much less affected. The predicted decrease in SWA for a realistic line roughness of σ=2.5 nm (rms) is as high as 5° and, thus, resolves the observed discrepancy between scatterometry and AFM results.

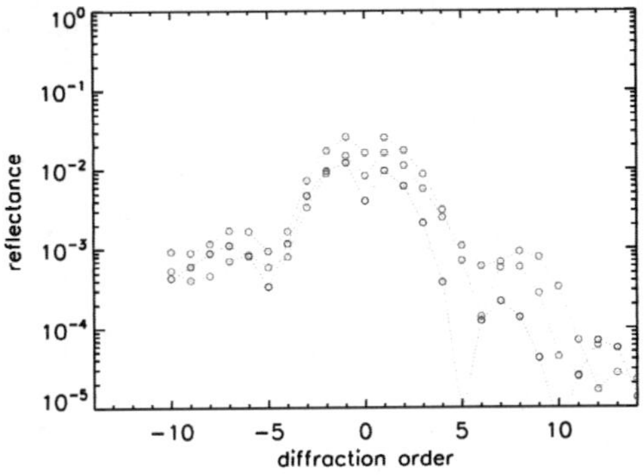

Figure 6. Diffraction intensities for semi-dense nominally 180 nm bright lines at 720 nm pitch. Data for 3 EUV wavelengths at a 6° angle of incidence are shown (green: 13.4 nm, red: 13.65 nm, and blue: 13.92 nm). The lines shown are only to guide the eye.

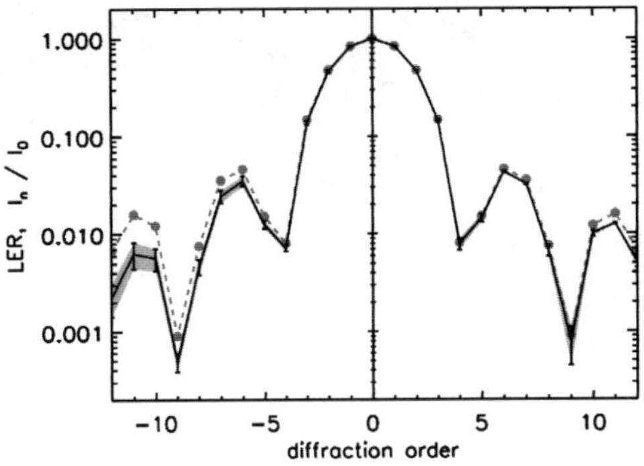

Figure 7. Diffraction pattern (Fraunhofer approximation, red bullets) and the expected intensity pattern with its standard deviation for each diffraction order: LER in the left half, LWR on the right, 165 nm bright lines at 720 nm pitch, σ = 10 nm, number of lines N = 100.

5. Conclusion

At PTB, scatterometry with angular resolution using EUV radiation is used for the characterization of EUV multilayer mirrors. The PSD can be measured up to spatial wavelengths of 10 μm and high angular frequency modulations in the scattered intensity corresponding to a 100 μm feature size can be resolved. On the other hand, scatterometry using discrete diffraction orders from periodic structures is used to determine geometrical structures from inverse modelling. In a detailed investigation of the measurement uncertainties, it is shown that structural imperfections, like edge roughness, have the highest impact on the reconstruction of the geometrical parameters by angular resolved scatterometry. An improved modelling of roughness is, therefore, of the utmost importance. A first step to combine the statistical approach with structure reconstruction by FEM calculations is presented.

References

[1] Stover J C, "Optical scattering," SPIE Optical Engineering Press, Bellingham, Washington, USA (1995)
[2] Raymond C 2005 *AIP Conf. Proc.* **788** 394-402
[3] Stork W, Streibl N, Heidner H and Kipfer P 1991 *Opt. Lett.* **16** 1921-3
[4] Trenkler J et al. 2002 PXRMS'02; Chamonix, Talk 5.3
[5] Scholze F and Laubis C 2008 *Proc. SPIE* **6792** 67920U
[6] Scholze F, Laubis C, Ulm G, Dersch U, Pomplun J, Burger S and Schmidt F 2008 *Proc. SPIE* **6921** 69213R
[7] Pomplun J, Burger S, Schmidt F, Scholze F, Laubis C and Dersch U 2008 *Proc. SPIE* **7028** 70280P
[8] Kato A and Scholze F 2010 *Appl. Opt.* **49** 6102-10
[9] Scholze F et al. 2006 *Proc. SPIE* **6151** 615137
[10] Klein R, Laubis C, Müller R, Scholze F and Ulm G 2006 *Microelectr. Eng.* **83** 707-9
[11] Scholze F et al. 2001 *Proc. SPIE*, **4344** 402-13
[12] Lowisch M, Kuerz P, Mann H J, Natt O and Thuering B 2010 *Proc. SPIE* **7636** 763603
[13] Gross H, Rathsfeld A, Scholze F and Bär M 2009 *Meas. Sci. Technol.* **20** 105102

Journal of Physics: Conference Series **311** (2011) 012007 doi:10.1088/1742-6596/311/1/012007

Machine integrated optical measurement of honed surfaces in presence of cooling lubricant

R. Schmitt[1,2], N. König[1,2] and H. Zheng[1]

[1]Laboratory for Machine Tools and Production Engineering of RWTH Aachen University, Steinbachstr. 19, 52074 Aachen, Germany

[2]Fraunhofer Institute for Production Technology IPT, Steinbachstr. 17, 52074 Aachen, Germany

E-mail: n.koenig@wzl.rwth-aachen.de

Abstract. The measurement of honed surfaces is one of the most important tasks in tribology. Although many established techniques exist for texture characterization, such as SEM, tactile stylus or white-light interferometry, none of them is suited for a machine integrated measurement. Harsh conditions such as the presence of cooling lubricant or vibrations prohibit the use of commercial sensors inside a honing machine. Instead, machined engine blocks need time-consuming cleaning and preparation while taken out of the production line for inspection. A full inspection of all produced parts is hardly possible this way. Within this paper, an approach for a machine-integrated measurement is presented, which makes use of optical sensors for texture profiling. The cooling lubricant here serves as immersion medium. The results of test measurements with a chromatic-confocal sensor and a fiber-optical low-coherence interferometer show the potential of both measuring principles for our approach. Cooling lubricant temperature and flow, scanning speed and measurement frequency have been varied in the tests. The sensor with best performance will later be chosen for machine integration.

1. Introduction

Within the production chain of combustion engines, the honing process for the manufacturing of tribological surface structures within cylinder liners is crucial for the efficiency and oil consumption of the engine. The quality inspection of honed textures today is mostly realized with tactile stylus, fax image analysis [1] or scanning electron microscope (SEM) [2], while tactile probing is the first choice for the production-related evaluation. For all of the existing measuring principles, the separation of the engine block from the production line and additional washing is needed, which is time consuming. While a complete inspection of all produced parts is impossible, the sample testing causes another problem: if the time delay to prepare and measure a cylinder liner is too long, a lot of engine blocks can be manufactured with possible defects in the meantime. Therefore, a machine integrated measuring method is highly demanded, which enables the fast inspection of the honed texture.

2. Optical measurement system

2.1. Conceptual design of the measurement set-up

As introduced above, the cooling lubricant has to be cleaned before the tactile measurement of the surface. Within this paper, an innovative approach with optical sensors is presented, in

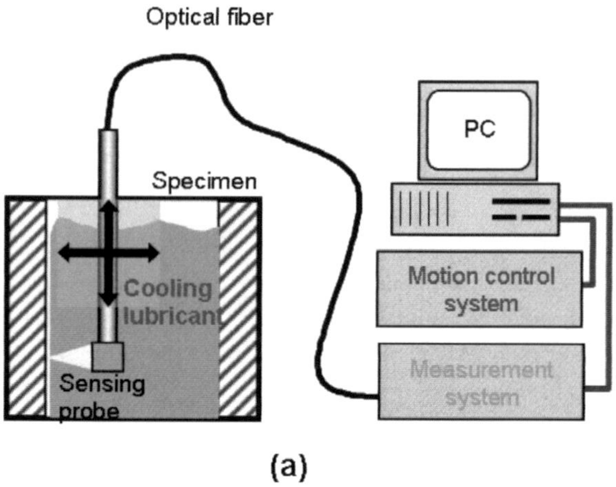

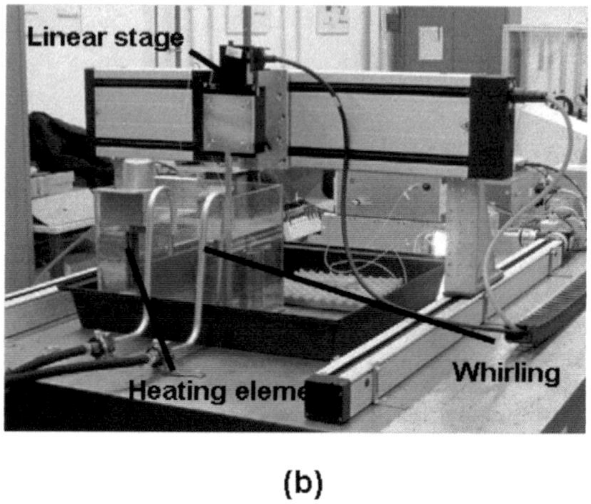

Figure 1. Optical measurement system (a) and test bench for sensor evaluation (b)

which the cooling lubricant serves as immersion medium. As shown in the conceptual set-up in Fig. 1(a), the cylinder of the engine block is fully filled with the cooling lubricant. The sensing probe is totally immersed into the cooling lubricant during the measurement. The measurement signals are then transferred through a fiber to the measuring system, which also contains a positioning module and synchronization module.

Since the measuring system will be later integrated into a honing machine to conduct inline surface measurement, a test bench has been built up in order to simulate the production conditions during the honing process. Different factors, e.g. rough and fine surface, change of lubricant temperature and whirling, scanning speed etc., can be analyzed for selected optical sensors. The test bench is shown in Fig. 1(b).

2.2. Optical sensors

Due to the fine profile (average roughness Ra smaller than 0.5 µm) and small diameter (approx. 85 mm) of the cylinders of the engine block, the optical sensors should be sensitive enough and should be able to be miniaturized to enable a measurement in the cylinder. Two fiber based sensors, a chromatic-confocal sensor and a fiber-optical low-coherence interferometer, were selected based on their measuring principles.

The chromatic-confocal principle makes use of the high chromatic aberration of the sensor focussing lens. The measured distance is spectrally encoded and the spectrum is then analyzed by a spectrometer [3]. An optical fiber, which delivers the optical signal from the sensing probe to the spectrometer, serves as a confocal aperture. For our tests, a chromatic-confocal sensing probe with a diameter of 10 mm, a measuring range of 100 µm and a high numerical aperture (NA) of 0.9 for the measurement in cooling lubricant has been specially designed and built by Precitec Optronik GmbH of Rodgau, Germany.

The second sensor under test is a fiber-optical low-coherence interferometer, which has been developed at Fraunhofer Institute for Production Technology IPT and is now provided by fionec GmbH of Aachen, Germany. The sensing principle is based on low-coherence interferometry and is described in [4]. The sensing probe is built in all-fiber design, but has been integrated

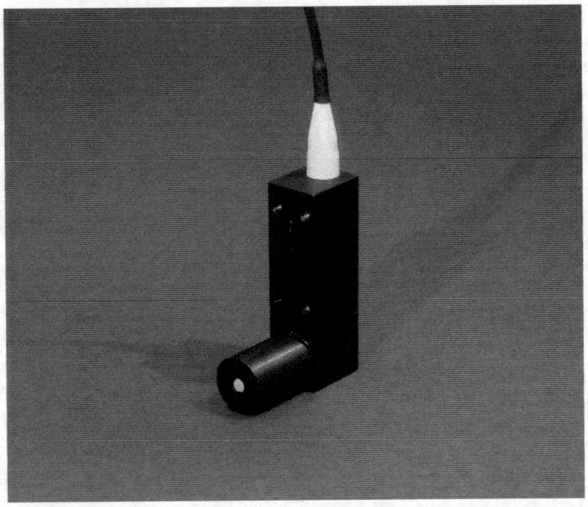

Figure 2. Fiber optic distance sensor for machine integration

into a similar shaped probe housing (Fig. 2) as the chromatic-confocal sensor for compatibility reasons.

3. Evaluation measurements

To evaluate the performance of both sensors, a series of test measurements have been carried out, in which different objects have been measured with both sensors in cooling lubricant as follows:

3.1. Calibration

Prior to roughness measurements, the distance between the surface and probe had to be calibrated in cooling lubricant, since the refractive index of the lubricant serves as immersion medium and affects the optical path length of the measuring beam. Since the measuring processes of both sensors are similar to a tactile stylus profiler, a depth-setting standard has been used for calibration according to ISO 5436-1 type A 1 [5]. For calibration the length of measurement must be at least three times the width W of the groove. To discard data losses or batwings at the groove edges, a range of $\pm W/3$ is skipped on each side of the groove. With the remaining measurement data, ranging from point 1 to 2, 3 to 4, 5 to 6 in Fig. 3a, the groove depth D can be determined.

To calibrate the measurement scale and to compensate for a possible nonlinearity, a depth setting standard with two grooves, of which one is 20 µm and the other 50 µm deep, has been used (Fig. 3b). The depth setting standard was calibrated with a white-light interferometer Veeco NT1100. Since the refractive index of the cooling lubricant can change with many factors such as temperature, cleanliness or type of cooling lubricant, a calibration with the depth setting standard has been conducted for different measuring setups.

3.2. Test on roughness standards

Although both sensors have been characterized for roughness measurements in dry conditions [6], different roughness standards with Rz values of 3.2 µm, 10 µm (Fig. 4) and 20 µm were measured in the test bench filled with cooling lubricant. Different measuring frequencies (constant sampling interval) and setups were tested.

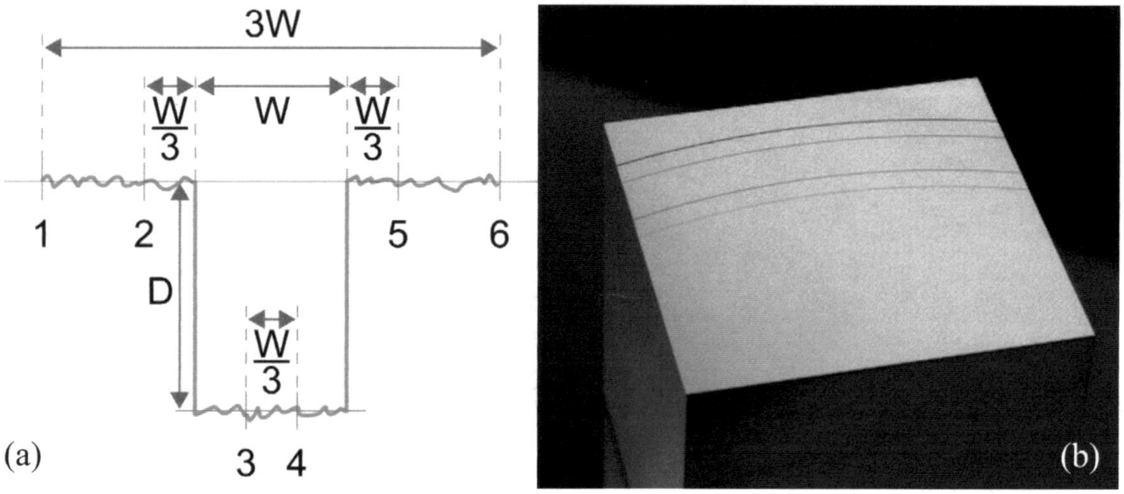

Figure 3. a) profile section of the depth-setting standard, b) diamond turned depth setting standard

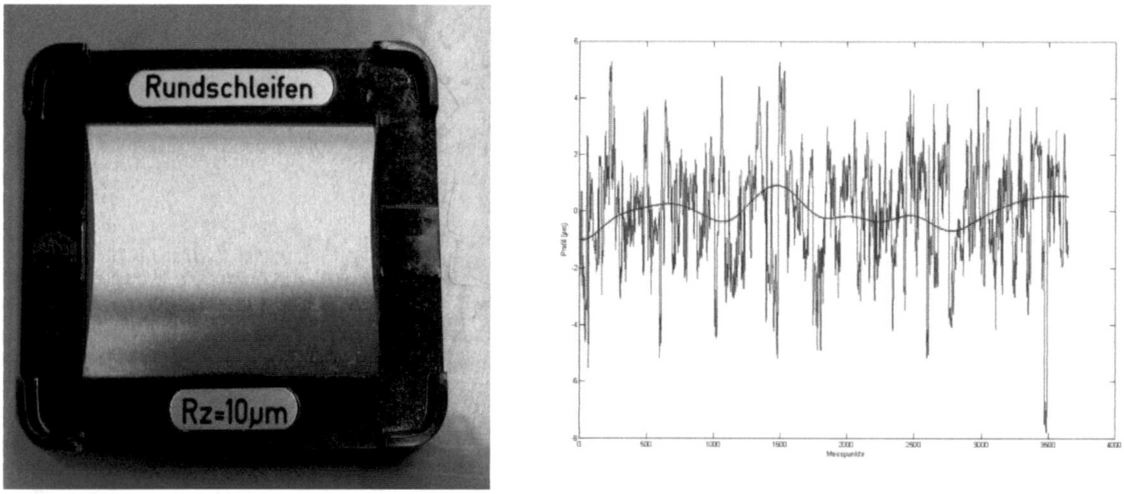

Figure 4. Roughness standard with Rz of 10 µm and profile measured with fiber-optical distance sensor

Both sensors have proven capable for roughness measurement in the cooling lubricant under production conditions. The results show that the deviations from the calibrated value for the measured Rz values of both sensors are within ±15% and the deviations for measured Ra values are within ±6.6%.

3.3. Tests on honed cylinder surfaces

A series of test measurements on honed cylinder surfaces has been made with both optical sensors. Different factors, which may influence the measurement results, e.g. temperature, whirling, lateral resolution, and so on, have been varied according to a design of experiments test chart (see table 1).

Table 1. Experimental design of tests on cylinder surfaces

Factors	Factor level 1 (-)	Factor level 2 (+)
(A) Lateral resolution	1 µm spacing	1,5 µm spacing
(B) Measuring frequency	low (1 kHz)	high (2 kHz)
(C) Whirling	Off	On
(D) Tilting	low (< 0,02°)	high (> 0,1°)
(E) Temperature	21°C	39°C

Altogether 64 measurements have been conducted with both sensors. The results indicate that the optical measurements are mostly affected by the temperature effect, which causes a change in the density of the cooling lubricant and, related to that, a change of its refractive index. In real production processes, the lubricant temperature is actively cooled and kept constant within a few degrees of Kelvin, so that the temperature effects on the measurement are not considered to be problematical. Compared to results, that have been acquired with tactile stylus measurements, the expanded uncertainty for Ra (using a coverage factor of 2), is approximately 80 nm for both optical sensors. Ligth scattering methods, which have emerged in recent years, could be an alternative for a quick inline assessment within cooling lubricant. For a reliable determination of the feasibility, extensive optical simulation of the scattering behavior of the honed structure in cooling lubricant is crucial.

4. Conclusion
Within this paper we presented a new approach for a machine integrated inspection of honed surfaces. The used optical sensors provide the potential for the fast inline assessment of the honed texture. Different evaluation measurements show that the sensors are capable for use in cooling lubricant. In a next step, further evaluation measurements will be performed with cooling lubricant emulsion. One sensor will be chosen for integration into a honing machine, which can later be inserted into a production line for the automated quality inspection and closed-loop control of the honing process.

Acknowledgments
The presented results have arisen from the national research project "ZOOM", which is kindly supported by the German Ministry of Education and Science (BMBF) and the VDI (FKZ 13N10294).

References
[1] Geus D and Stiebler M 2008 *Meas. Sci. Technol.* **19** 064004
[2] Willis E 1986 *Wear* **109** 351-66
[3] Tiziani H J and Uhde H M 1994 *Appl. Opt.* **33** 1838-43
[4] Depiereux F, Lehmann P, Pfeifer T and Schmitt R 2007 *Appl. Opt.* **46** 3425-31
[5] ISO 5436-1:2000 *Geometrical Product Specifications (GPS) - Surface texture: Profile method; Measurement standards - Part 1: Material measures*
[6] Schmitt R, König N and de Araújo E M 2009 *Proc. of SPIE (Munich)* vol 7389, ed P Lehmann (Bellingham: SPIE) p 7389-14

Journal of Physics: Conference Series **311** (2011) 012008 doi:10.1088/1742-6596/311/1/012008

Effect of measurement conditions on three-dimensional roughness values, and development of measurement standard

A Fabre[1,2], S Raynaud[3], B Brenier[2],

[2] Arts et Métiers ParisTech, MécaSurf Laboratory, 2, Cours des Arts et Métiers, 13617 Aix-en-Provence, France

[3] INSA Lyon, MIP2 Laboratory, 27 Avenue Jean Capelle, Bat Jacquard, 69100 Villeurbanne

E-mail: agnes.fabre@ensam.eu

Abstract. Friction or corrosion behaviour, fatigue lifetime for mechanical components are influenced by their boundary and subsurface properties. The surface integrity is studied on mechanical component in order to improve the service behaviour of them. Roughness is one of the main geometrical properties, which is to be qualified and quantified. Components can be obtained using a complex process: forming, machining and treatment can be combined to realize parts with complex shape. Then, three-dimensional roughness is needed to characterize these parts with complex shape and textured surface. With contact or non-contact measurements (contact stylus, confocal microprobe, interferometer), three-dimensional roughness is quantified using the calculation of pertinent parameters defined by the international standard PR EN ISO 25178-2:2008. An analysis will identify the influence of measurement conditions on three-dimensional parameters. The purpose of this study is to analyse the variation of roughness results using contact stylus or optical apparatus. The second aim of this work is to develop a measurement standard well adapted to qualify the contact and non-contact apparatus.

1. Introduction

Some characterizations of roughness are investigated with different apparatus. Three-dimensional (3D) roughness was characterized on two samples (a ring and a tube) and the first standard of measurement developed in this current study [1,2,3]. The influences of measurement conditions on roughness parameters are analysed, for contact and for non-contact measurements: in terms of step of scanning and size of area [4]. Roughness values are compared after two measurements on the same place inside the tube. The nomenclature is given on the table1. These measurements were performed using a stylus contact profilometer and a confocal sensor. The standard of measurement is created with a specific geometry to enable qualification of roughness with different instrument.

Table 1. Nomenclature.

(X, Y, Z)	Lateral coordinates of point on the surface and vertical measurement coordinate on the surface
(A, B)	Big radius of ring, about 70 mm and respectively, small radius of ring, about 13 mm
(C, D)	Radius of the tube, about 10 mm, and respectively, thickness of the tube, about 1mm
St	Maximal height, µm

[1] Arts et Métiers ParisTech, MécaSurf Laboratory, 2, Cours des Arts et Métiers, 13617 Aix-en-Provence, France

Sz	Ten points height of surface, µm
Sds	Density of summits, mm^{-2}
Sq	Root mean square deviation, µm
Ssc	Arithmetic mean peak curvature, μm^{-1}
Ssk	Skewness, µm
Sku	Kurtosis, µm
Sdq	Root mean square slope, µm/µm
Sdr	Developed interfacial area ratio, %

2. Influences of measurement conditions

The influences of measurement conditions are analysed after contact and non-contact measurement. All the apparatus used are equipped with a stylus or a point sensor. The relevant techniques used in this current study qualify Z, the altitude position of points on the surface characterized for each position (X,Y), with specific associated steps. The system of coordinate is established on the reference of the apparatus, associated to the motion in X and Y directions.

2.1. Contact measurements

The first sample is a ring. The figure 1 shows the geometry of this sample, with a radius A of 70 mm and with a radius B of 13 mm. Some measurements were done on the torus of the ring using the contact device SURFASCAN 3D to analyse the variation of the 3D surface characterization. The vertical resolution of the inductive transducer is 4 nm. The radius of the stylus used is about 2 µm and the angle of the edge is about 60°. The measurements were done on the torus on an area about 8 mm × 4 mm. These measurements were done with a step in X direction about 4 µm and in Y direction about 4.4 µm.

Figure 1: Measurement on ring with contact 3D apparatus.

The least mean square plane associated to the measurement points is characterized. Then, the correction of the defect of parallelism between the surface and the reference of measurement is corrected using the appropriate rotations. The effect of F-Filter was analysed with different length of spline. The shape involved was distorted. Then, no F-Filter was used for treatment, but a cropping was done, in order to reduce the zone analysed (without perturbation by the shape). The L-filter, a 3D Gaussian regression filter with a cut-off about 0,8 mm is applied [5]. The S-filter, a 3D Gaussian regression filter with a cut-off about the double of the lower step was applied. No significant evolution on parameters are analysed with and without S-filtering. Then, the L-filter is used such as simplified and sufficient treatment combining with cropping. The roughness parameter values are obtained using the Surfascan software.The value of Sa, the arithmetic roughness is about 0,14 µm. Height parameters are calculated using the definition given on the papers referenced [2,3,4] for the Sa, Sds, Sq, Ssc, Ssk, Sku, Sdq and Sdr. The Sz and St parameters are defined only in the references [3,4]. For each parameter analysed, the normalized results are given in the figure 2. The reference values are that obtained on the more precise measurement corresponding to the measurement with the smaller steps. 2, 3, 5, 7 and 10 multiplies the smaller steps. The influence of the size of the area of the measurement is investigated; normalized results are given in the figure 3. The reference values are that obtained on the more precise measurement corresponding to the measurement with the bigger area about 4 mm x 4 mm. The smaller areas are squares with a length about 2 mm and 1 mm.

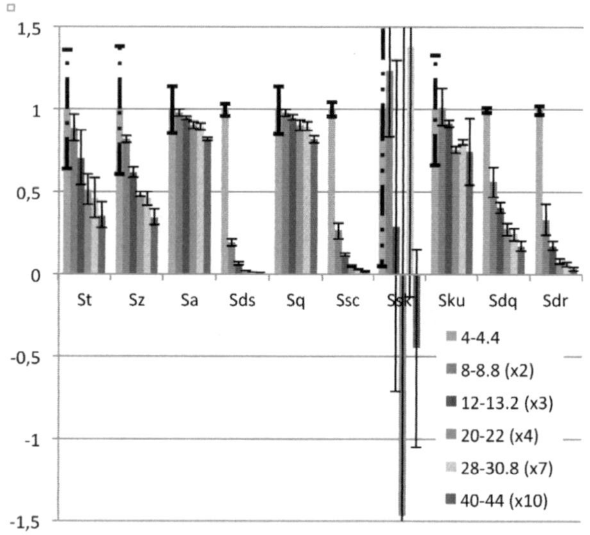

Figure 2: Influence of the steps acquisition on the normalized values of 3D roughness parameters, done using contact apparatus.

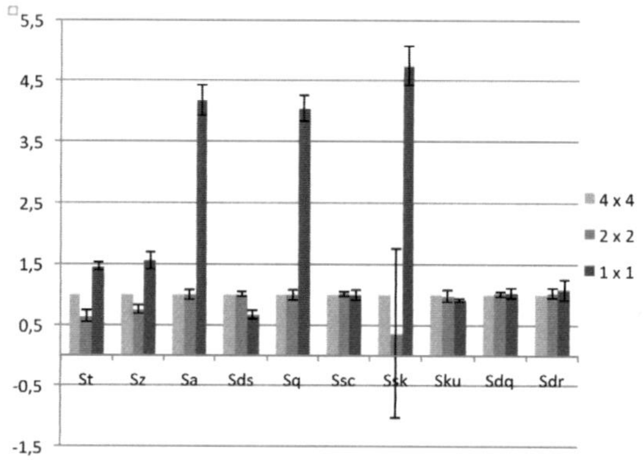

Figure 3: Influence of the size of the acquisition area on the normalized values of 3D-Roughness parameters, measurement done using contact apparatus.

2.2. Non contact measurements

The second sample is a tube. The figure 3 shows the geometry of this sample, with a radius C of 10 mm and with a thickness D of 1 mm. Some measurements were done using the confocal chromatic sensor on the device Altimet 3D to analyse the variation of the 3D surface characterization. The vertical resolution of the optical pen is 3 nm. The lateral resolution of the optical pen is 3.3 μm. The measurements were done inside the tube, on an area about 2 mm^2, with a step about 10 μm in the both directions X and Y.

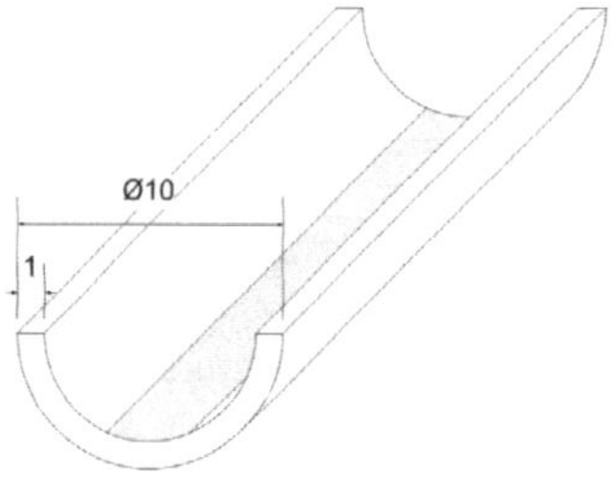

Figure 4: Measurement on a tube with a 3D confocal apparatus.

The shape was cut-off using a F-Filter with length of spline about 0,8 mm. The L-filter, a 3D Gaussian regression filter with a cut-off about 0,8 mm is applied [5]. The S-filter, a 3D Gaussian regression filter with a cut-off about the double of the step was applied. No significant evolution on parameters are analysed with and without S-filtering. Then, the results are obtained by combining F-filtering, L-filtering and rotation of the least mean square plane. The roughness parameter values are obtained using the Surfascan software. The value of Sa, the arithmetic roughness is about 0,34 μm. For each parameter analysed, the normalized results are given in the figure 5. The reference values are that obtained on the more precise measurement corresponding to the measurement with the smaller steps (10 μm). 2, 3, 4 and 6 multiplies the smaller steps.

The influence of the size of the area of the measurement is investigated; normalized results are given in the figure 6. The reference values are that obtained on the more precise measurement corresponding to the measurement with the bigger area about 1 mm x 2 mm. The smaller areas are squares with a length about 1 mm.

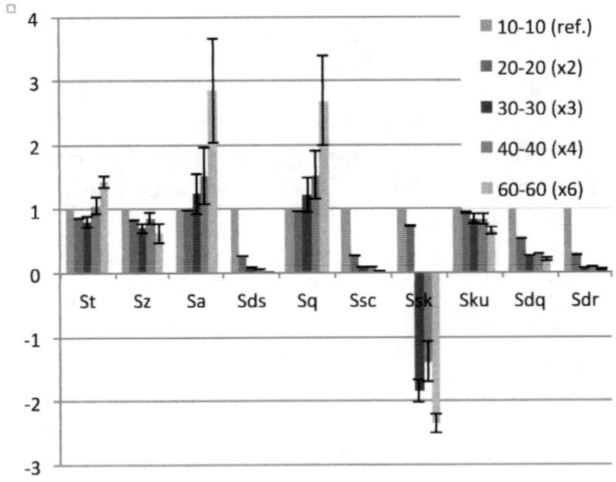

Figure 5: Influence of the steps acquisition on the normalized values of 3D-Roughness parameters, measurement done using non-contact apparatus

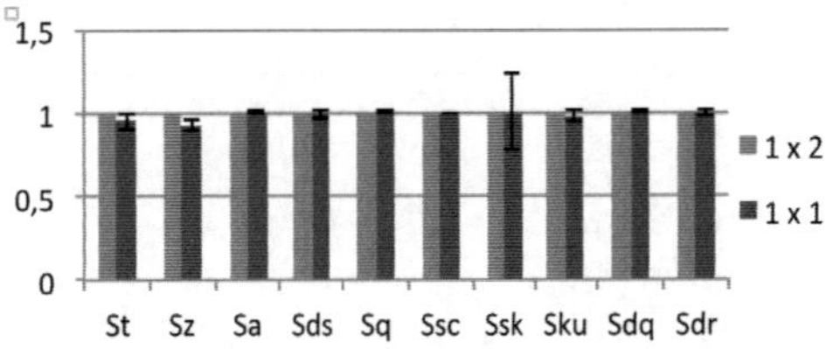

Figure 6: Influence of the size of the acquisition area on the normalized values of 3D-Roughness parameters measurement done using non-contact apparatus.

2.3. *Influences on results*

On the results given in the figure 2 and the figure 5, only five of the ten parameters are strongly influenced by the step when measurement is done with contact profilometer and confocal sensor. Firstly, the value of the skewness, Ssk, seems to be very instable. The values of the spacing parameter, Sds and the hybrid parameters Ssc, Sdq decrease more than 40 % when the step is divided by two; Sdr decrease more than 60 %. In the same case, the values of local amplitude parameters St and Sz decrease no more than 20 %. For the two parameters Sa and Sq, the behaviour is not the same for the ring and the tube. The results given on the ring sample with the contact apparatus produce almost similar values of these two parameters. Less than 20% of decrease can be observed on the Sa and the Sq values, when the step is divided by ten. The increase of the Sa and the Sq values are about 20 % in the case of a non-contact measurement on the tube, when the steps are divided by three.

On the results given in figure 3 and figure 5, only three parameters are strongly influenced by the size of the area scanned in the case of contact or non-contact measurements when the size is divided by two or fourth. The values of the St, Sz and Ssk are significantly affected. The variation of Sa and Sq values are less than 1.1 % in the case of a non-contact measurement on the tube. No more than 1% of variation can be observed on the Sa and Sq values, when the area is divided by fourth, for the results given on the ring sample with the contact apparatus. These values are strongly disturbed when the size is divided by sixteen.

2.4. Optimal method of measurement with accuracy

Previous analyses clearly indicate that the results on Sds, Ssc, Sdq and Sdr values are very influenced by the step of the scanning. The area size is not so influent. It will be interesting to develop an optimal method in order to quickly qualify the roughness. A first measurement with smallest steps, on a small area, can allow identifying the stable parameters: for example, such as Sa and Sq, for the results obtained on the ring. Then, the optimal step of scanning can be used to perform a series of measurements. In every case, measurement with the smallest step will enable to determine precisely hybrid parameters Ssc, Sdq, Sdr and Sds.

2.5. Influence of the apparatus

The tube defined on figure 2 was measured with a step of 10 µm in a previous study [6], with contact profilometer and confocal sensor. The conditions of measurements are given on the table 2. In the figure 7, the values of some 3D-Roughness parameters, calculated after contact measurement, are used to define the reference and there are compared with values obtained using confocal sensor. Variations can be observed on all the values characterized by the two techniques. The values of average parameters, as Sa and Sq, present a variation more than 40 %. Then, it seems to be necessary to measurement a standard, in order to precise the values of roughness.

Table 2. Conditions of contact and non-contact measurement inside the tube.

Measurement conditions	Contact	Confocal
Area, mm^2	4	2
Step in Y and Y directions, µm	10	10
Stylus radius or spot size, µm	2	3,3

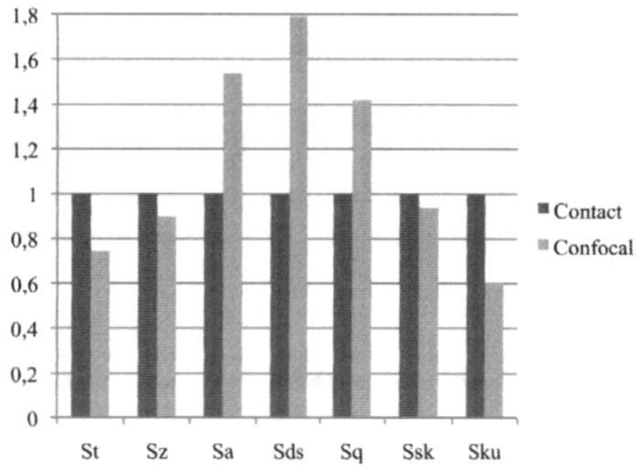

Figure 7: Influence of the apparatus values of 3D-Roughness parameters, measurement done using contact profilometer and non-contact confocal instruments

3. Development of a standard of measurement

Some standards are developed to verify the roughness but on plane samples. Others standards can be used to verify the capability of the apparatus only at the scale of macrogeometry [7]. In fact, all the previous analysis conducted us to develop a specific measurement standard of roughness with a complex shape. Then, the identification of properties needed for this part is done. This standard must allow quantifying roughness using different instruments. The geometry is defined with a complex and specific shape: with concave and convex curvature. A measuring range about three hundreds micrometers is selected: optical apparatus, as interferometer or confocal sensor must enable to measurement this part. Then, dimensions are computed in order to allow the measurement on a maximal area about 4 mm x 4 mm. The first sample was turned on a bar of 35NiCrMo16, with hardness about 40 HRC. Turning process is chosen in order to create different roughness levels due to different feedrate values. The drafting is given in figure 8(a).

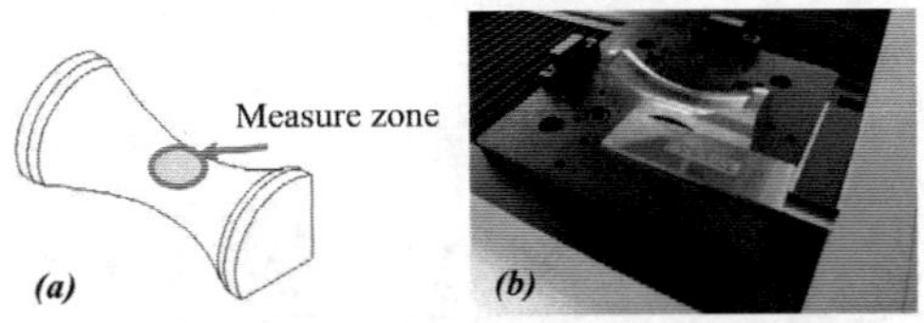

Figure 8: Measurement standard (a) drafting, (b) position in measurement statement.

A protocol of manipulation was established when the first measurements were performed with this standard. The figure 8(a) indicates the specific position of the measurement, in the middle of the sample. A method is developed to allow a precise positioning of the sample on the apparatus plate: the axis of the sample needs to be parallel to the X motion of the table (figure 8(b)). After the first measurement on the standard, an improvement of this component is needed. The standard presents a shape defect. Some investigations with mill-turn machining will be conducted in next months in order to reduce this defect. On the other hand, the protocol must be adjusted in order to precise the treatment of measurement data. First analysis had shown a significant influence of the shape filters on the roughness results.

4. Conclusion

A standard of roughness is developed; a protocol of measurement is defined in order to quantify the roughness parameters. In the next year, a new standard will be used to investigate with a campaign of inter-comparison on different instruments. The results obtained will improve the knowledge of influence factors on values of roughness parameters. Maybe specific and optimal method of characterization will be defined.

5. Acknowledgements

Thanks to Y. Cao, A. Soller, R. Baloche, student engineers, L. Remandet, from CTDeC, N. Boucher, F. Ville and P. Sainsot, from LaMCoS of INSA of Lyon for their contribution in this study.

6. References

[1] Norme XP E05-030-1, Etat de surface : Surfacique - Partie 1 : Termes et définitions et paramètres surfaciques, Décembre 2003.

[2] PR EN ISO 25178-2, Geometrical product specifications-Surface texture: Areal-Part2: Terms, definitions and texture parameters, april 2008.

[3] K. J. Stout, P. J. Sullivan, W. P. Dong, E. Mainsah, N. Luo, T. Mathia, and H. Zahouani - *The Development of Methods for the Characterization of Roughness in Three Dimensions*, Pub. No. EUR 15178 EN, University of Birmingham, 1993.

[4] R. Ohlsson, B. G. Rosen, J. Westberg - The interrelationship of 3D Surface, characterisation techniques with standardised 2D techniques, L. Blunt L, X. Jiang, *Advanced Techniques for Assessment Surface Topography. Development of a Basis for 3D Surface Texture Standards Surfstand*, Ed. Kogan page science, UK, 2003, pp. 197-220.

[5] S. Brinkman, H. Bodschwinna, - Advanced Gaussian filters, in: L. Blunt L, X. Jiang, *Advanced Techniques for Assessment Surface Topography, Development of a Basis for 3D Surface Texture Standards Surfstand*, Ed. Kogan page science, UK, 2003, pp. 63-89.

[6] S. Raynaud, A. Soller, A. Fabre; Intercomparaison pour la qualification d'état de surfacique, in Congrès International de Métrologie, Paris, France, 22-25 juin, 2009.

[7] J. F. Ville, Calibration procedures for stylus and Optical instrumentation, in: L. Blunt L, X. Jiang, *Advanced Techniques for Assessment Surface Topography, Development of a Basis for 3D Surface Texture Standards Surfstand*, Ed. Kogan page science, UK, 2003, pp. 119-174.

Development of measurement standards for verifying functional performance of surface texture measuring instruments

A Fujii[1], H Suzuki[2] and K Yanagi[3]

[1]Life and Industrial Product Development Department Olympus Corporation, 2951 Ishikawa-machi, Hachiouji-shi, Tokyo, Japan

[2]Industrial Marketing & Planning Department Olympus Corporation, Shinjyuku Monolith, 3-1 Nishi-Shinjyuku 2-chome, Tokyo, Japan

[3]Department of Mechanical Engineering ,Nagaoka University of Technology, 1603-1 Kamitomioka-machi, Nagaoka-shi, Niigata, Japan

E-mail: a_fujii@ot.olympus.co.jp

Abstract. A new measurement standard is proposed for verifying overall functional performance of surface texture measuring instruments. Its surface is composed of sinusoidal surface waveforms of chirp signals along horizontal cross sections of the material measure. One of the notable features is that the amplitude of each cycle in the chirp signal form is geometrically modulated so that the maximum slope is kept constant. The maximum slope of the chirp-like signal is gradually decreased according to movement in the lateral direction. We fabricated the measurement standard by FIB processing, and it was calibrated by AFM. We tried to evaluate the functional performance of Laser Scanning Microscope by this standard in terms of amplitude response with varying slope angles. As a result, it was concluded that the proposed standard can easily evaluate the performance of surface texture measuring instruments.

1. Introduction

The Design of the surface texture and its realization are of great importance in advanced industrial products. Certain measuring instruments for quality control are necessary at the shipment stage. Consequently, many types of surface texture measuring instruments have appeared on the market. Among them, the optical measurement method occupies an important position because it can achieve the non-contact and fast measurement even in the manufacturing line. Moreover, the areal (3D) measurement instead of a simple profile (2D) measurement is increasingly demanded. Although the optical surface texture measuring instruments have been gradually popularized, they are hardly authorized so far. This is partly because of being based on different measuring principles. To make matters worse, they are not always correlated each other [1, 2]. We previously reported that this causes different response curves against surface wavelengths in every measuring instrument [3].

Since the instrument response curve represents its overall measuring performance, it is obligatory for instrument users to obtain response curves in advance and understand the capability for measuring the surface texture. The working group of ISO/TC213 are being discussed the measurement standards for verifying the instrument measuring performance. Those measurement standards are designed mainly to calibrate the measuring axes of instruments, e.g. X, Y and Z, respectively. However, no measurement standards for evaluating response curves are presented in the working draft. Although the irregular random topography standard derived from a stochastic process is proposed [4], it can not be used for verifying the instrument response characteristic but for checking the overall measuring performance through surface roughness parameter values.

In this paper, we propose a complementary measurement standard for verifying overall measuring performance of areal surface texture measuring instruments in conjunction with response curves against surface wavelengths.

2. Design procedure for measurement standard

2.1. Existing condition

Step height standards and sinusoidal waveform standards are commercially available. In our previous study [3], a number of sinusoidal waveform standards in the longer surface wavelength range and 1-D grating (grid with horizontal spacing) artifacts in the shorter range were utilized for the measurements of response curves of some surface texture measuring instruments on the market. In order to adequately evaluate the instrument response curves, a series of sinusoidal waveform measurement standards composed of various wavelengths and amplitudes are needed. In addition, for the optical instruments, the local slope angle of the facet is an influence factor on the response curve of the instrument. A combination of wavelengths and amplitudes is design parameters to keep the slope angle constant. Then, strongly demanded is just a measurement standard with the sophisticated surface waveform which is similar to the chirp signal.

2.2. Mathematical Model

From the above-mentioned demand, we developed the sinusoidal waveform from the pure chirp signal into the chirp-like waveform with changing amplitudes. A mathematical rule was applied to the processing, that any cosine curves can be continuously connected each other at the point where the derivative is zero.

Let T_n be the surface wavelength to be connected, where n=1, 2, 3, ... N. To cover a wide band of wavelengths, the wavelength T_n to be connected is changed in geometric progression as expressed by equation (1).

$$T_{n+1} = aT_n \tag{1}$$

Let $f_n(x, y)$ be the one cycle of the cosine wave term of the wavelength T_n, and if its amplitude is normalized to 1, equation (2) is given as follows.

$$f_n(x,y) = \begin{cases} 0 & \left(x < \sum_{i=1}^{n-1} Ti\right) \\ \frac{1}{2}\left\{\cos\left[\frac{2\pi}{T_n}\left(x - \sum_{i=1}^{n-1} T_i\right)\right] - 1\right\} & \left(\sum_{i=1}^{n-1} Ti \le x \le \sum_{i=1}^{n} Ti\right) \\ 0 & \left(\sum_{i=1}^{n} Ti < x\right) \end{cases} \tag{2}$$

Summing up $f_{n-1}(x,y)$, $f_n(x,y)$ and $f_{n+1}(x,y)$ provides the curve, as shown in figure 1, where the cosine curves of different wavelengths are connected smoothly. In order to keep the slope angle constant, we have changed the amplitude of the cosine wave depending on wavelengths (Figure 1). This is the first notable feature of this measurement standard.

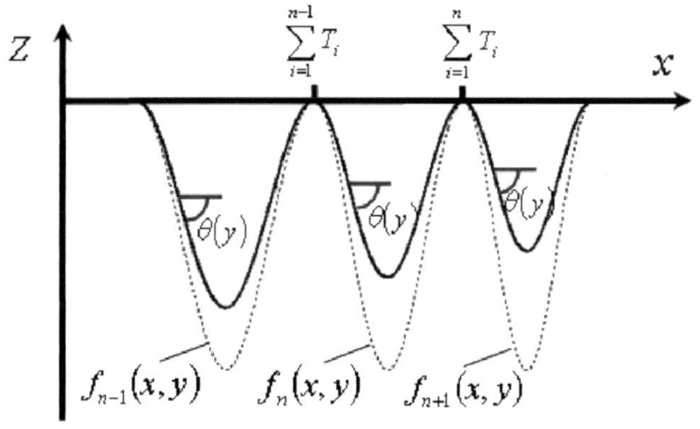

Figure 1. Chirp-like signal model in conjunction with constant slope angle.

If the amplitude of $f_n(x,y)$ is varied to make the local maximum slope angle in one cycle of the cosine wave to be $\theta(y)$, the topography data $Z(x,y)$ of the measurement standard summing up the varied amplitude of $f_n(x,y)$ is expressed by equation (3).

$$Z(x,y)=\sum_{n=1}^{N}\left[\frac{T_n \tan\theta(y)}{\pi} f_n(x,y)\right] \tag{3}$$

The reason why $\theta(y)$ is expressed by a function of y is to change the value of the local slope angle depending on the position of Y direction. This is the second notable feature of this measurement standard. Consequently, response properties including effects on the slope angle can be verified. Furthermore, to simplify the properties of measurement positions of samples when performing the cross-sectional analysis of measurement results, we changed $\theta(y)$ discretely (in step shape). $\theta(y)$ can be expressed by equation (4).

$$\theta(y)=\Delta\theta \,\text{floor}\left\{\frac{by+c}{\Delta\theta}\right\} \tag{4}$$

where, $\text{floor}(x)$ is the maximum integer below x against the actual number x, and $\Delta\theta$ is the step width of the discrete local slope angle.

2.3. Design Parameters

2.3.1. *Surface wavelength.* We determined design parameters for the measurement standard by selecting so-called Laser Scanning Microscope (LSM) as the target measuring instrument to be analyzed, and the following measuring conditions were set: objective lens 100x, NA 0.95, light source wavelength λ=0.405μm.

The optical cut-off wavelength (λ_{oc}) under these conditions is expressed by equation (5).

$$\lambda_{oc} = \frac{\lambda}{2NA} = \frac{0.405\mu m}{2 \times 0.95} \approx 0.21\mu m \tag{5}$$

Since λ_{oc} is a physical diffraction limit wavelength, the wavelength smaller than this cannot be observed. Therefore, we have set the minimum value of the wavelength T_n of the measurement standard as 0.2μm and the maximum value (=T_0) of T_n as T_0=2μm respectively, considering the field of view under the assumed measurement conditions. T_n is defined so that the parameter divides 1 decade from the wavelength 0.2μm to 2μm into 15 wavelengths geometrically. (T_0=2μm, n=15, a=0.85)

2.3.2. Local Slope Angle. In general, the measurable maximum slope of the stylus instruments is approximately 30 degrees. Accordingly, we consider the maximum slope angle of measurement standards for optical instruments is also 30 degrees. In this study, we set the large slope up to 45 degrees to check the limitation of measurable steep slope angle.

2.3.3. Consideration for processing. The unit pattern shown in figure 2 is derived from equation (1), (2), (3) and (4). In order to reduce the processing distortions caused by the elimination processing, this measurement standard model is structured by assigning 4 mirror patterns of the unit pattern symmetrically.

The size of patterning region is 24.3μm in X direction and 40μm in Y direction, which can be observed within one field view of 100x of LSM.

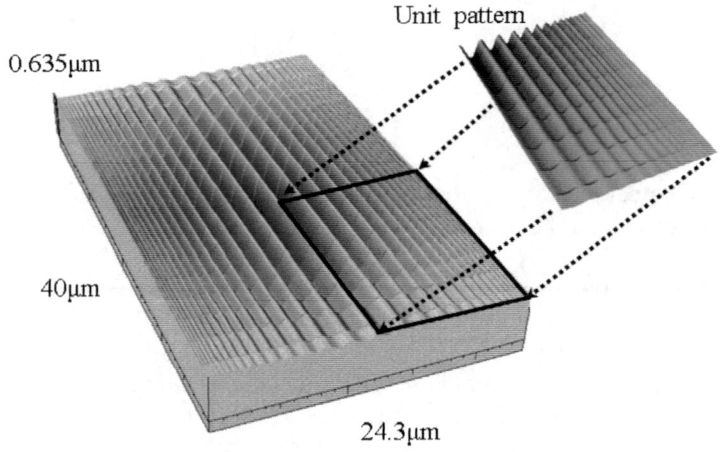

Figure 2. Measurement standard model designed for laser scanning microscope (LSM). T_0=2μm, a=0.85, n=15, b=-2, c=π/4 rad. =45 deg., $\Delta\theta$=π/36 rad. =5deg.

3. Fabrication

Measurement standard sample was fabricated by using FIB processing machine (Elionix Inc. EIP-5400). Figure 3 shows the optical microscope image of the fabricated sample surface. Figure 4 compares the designed profile and a measured profile by AFM. Two profiles are in good agreement. This means that the fabrication process was ideally performed.

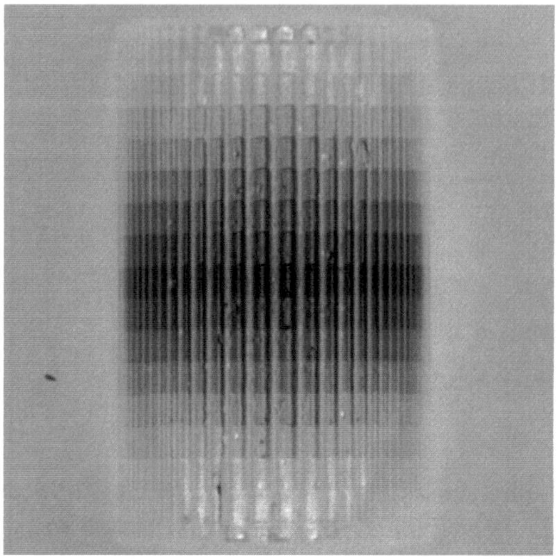

Figure 3. Microscope image of the fabricated sample surface.

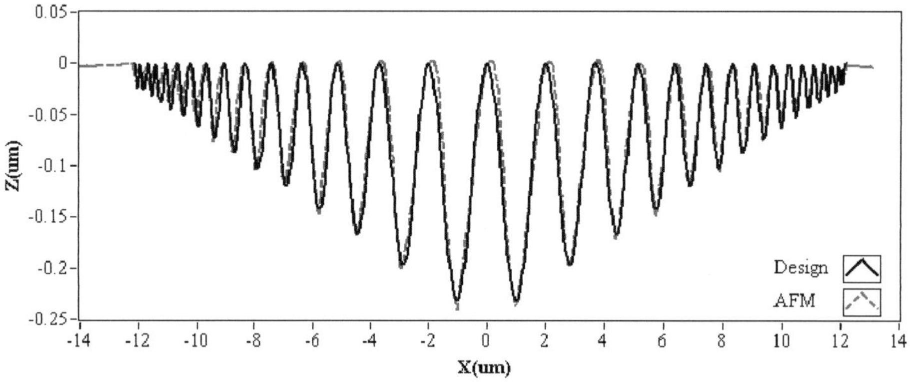

Figure 4. Comparison of profiles between design stage and measured by AFM (Maximum slope angle = 20 degrees).

4. Measurement results and discussion

Figure 5 shows the 3D topography measured by LSM (100x NA0.95). In this study, we defined the response as the ratio of LSM data to AFM data. Figure 6 plots LSM amplitude responses per each surface wavelength and per each slope angle. As shown in figure 5 and 6, if the slope angle is below 35 degrees, responses of LSM show the similar tendency, which is almost identical to results derived from our previous measurements conducted on numerous samples [3]. Now, we have made it possible to derive response properties such as figure 6 by only one measurement standard in only one time measurement.

As shown in figure 5, if the slope angle exceeds 35 degrees, it is recognized that spike noises increase gradually. We consider the phenomenon of spike noises recognized in the large slope is peculiar to all optical instruments because the similar case has been reported in the coherence scanning interferometry [6]. So, it is important to evaluate responses against the slope angle of optical measuring instruments accurately, and the measurement standard proposed by us will surely be its practical method.

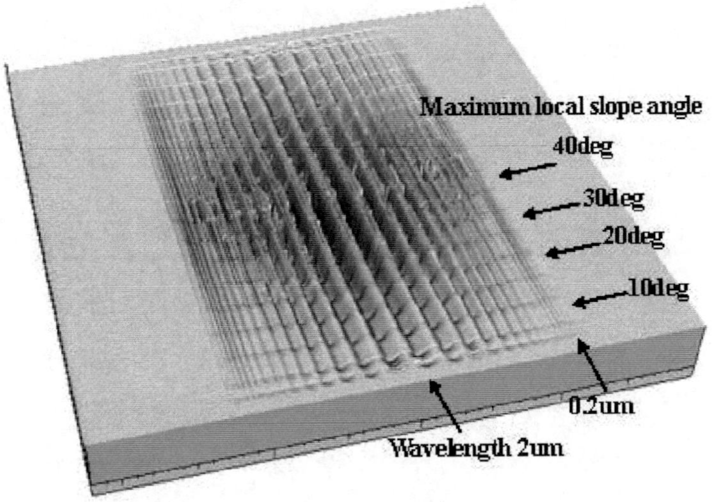

Figure 5. Bird's eye view of profiles measured by LSM.

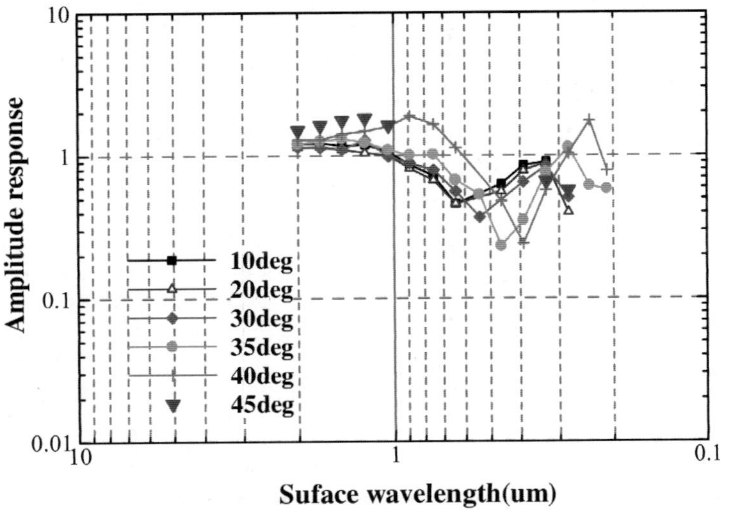

Figure 6. Response curves for the LSM at various slope angles (100x NA0.95).

5. Conclusion

- A new deterministic measurement standard was proposed to verify response properties and created by FIB processing.
- The response properties of Laser Scanning Microscope were evaluated by using the measurement standard.
- Response properties against various surface wavelengths and the effects of the local slope angle can be evaluated in one time measurement, which is proved to be the practical verification system.

Acknowledgments
The authors thank Mr.Y.Taguchi, Mr.T.Tsuchida, and Mr.M.Komazawa from Ellionix INC. for manufacturing the samples.

References

[1] Rhee H G, Vorburger T V, Lee J W and Fu J 2005 *Applied Optics* **44** 5919-27
[2] Leach R K, Haitjema H 2010 *Meas. Sci. Technol.* **21** 032001
[3] Fujii A, Suzuki H and Yanagi K, 2010 *Proc. ASPE Summer Topical Meeting on Precision Interferometric Metrology (Ashville,USA)* **49** pp91-96.
[4] Nemoto K, Yanagi K, Aketagawa M, Yoshida I, Uchidate M, Miyaguchi T and Maruyama H 2009 *Meas. Sci. Technol.* **20** 084023.
[5] Uchidate M, Yanagi K, Yoshida I, Shimizu T and Iwabuchi A 2009 *12th Int. Conf. on Metrology and Properties of Engineering Surfaces*, pp71-75
[6] Gao F, Leach R K, Petzing J, Coupland J M 2008 *Meas. Sci. Technol.* **19** 015303

Journal of Physics: Conference Series **311** (2011) 012010 doi:10.1088/1742-6596/311/1/012010

An expert system to characterize the surface morphological properties according to their functionalities

M Bigerelle[1], T Mathia[2], A Iost[3], T Correvits[4] and K Anselme[5]

[1] Laboratoire Roberval, UMR 6253, UTC/CNRS, UTC Centre de Recherches de Royallieu BP 20529, 60205 Compiègne France stol BS1 6BE, UK
[2] Laboratoire de Tribologie et Dynamique des Systèmes, UMR 5513, Ecole Centrale de Lyon, 36 Av Guy de Collongue, 69134 Ecully Cedex, France
[3] Laboratoire de Mécanique de Lille, UMR CNRS 8107, Arts et Métiers ParisTech - Lille, 8, boulevard Louis XIV 59046 Lille.
[4] Laboratoire de Métrologie. Arts et Métiers ParisTech, ENSAM, 8 boulevard Louis XIV, 59046 LILLE Cedex, France.
[5] Institut De Sciences Des Matériaux De Mulhouse, CNRS LRC 7228, 15, rue Jean Starcky, Université De Haute-Alsace, BP 2488, 68057 Mulhouse, France.

E-mail: maxence.bigerelle@utc.fr

Abstract. In this paper we propose a new methodology to characterize the morphological properties of a surface in relation with its functionality (tribological properties, surface coating adhesion, brightness, wettability…). We create a software based on experimental design and surface profile recording. Using an appropriate database structure, the roughness parameters are automatically computed at different scales. The surface files are saved in a hard disk directory and roughness parameters are computed at different scales. Finally, a statistical analysis system proposes the roughness parameter (or the pair of roughness parameters) that better describe(s) the functionality of the surface and the spatial scales at which the parameter(s) is (are) the more relevant.

1. Introduction

The precise characterization of machined surfaces roughness is of a major interest in many engineering industries due to its influence on the functionality of manufactured products. For obvious reasons of costs minimization, manufacturers are interested in developing reliable and simple control methodologies suitable for use in routine production environments, with a high degree of quantitativeness and data repeatability. The topographic method is by far the most implanted ones in surface quality assessment of metallurgical or mechanical products. The roughness of machined surfaces is of prime importance across a very wide spectrum of technical and scientific activities;

[1] To whom any correspondence should be addressed.

including not only tribologists and production engineers but also highway and aircraft engineers, hydrodynamicists and even bioengineers. Indeed, the morphology of machined surfaces often influences the properties that govern the application of the manufactured product. Consequently there still exists an increasing interest in developing reliable methodologies, suitable for quality control stage processes of surface products, in a manufacturing environment. Because of these various industrial and scientific interests, a proliferation of roughness parameters, possibly running into hundreds, has been triggered to describe the different kinds of surface morphology with regard to specific functions, properties or applications. In spite of this proliferation, termed by Whitehouse the "parameter rash", there is still no complete comprehensive account for the relevance of these roughness parameters. More, it becomes difficult to choose one parameter rather than another one. This probably comes from a lack of global methodology combined with the limits of the softwares used to characterize the surface morphology presently available on the market. In our opinion, the main objective of such a global methodology should be to determine quantitatively, and without preconception, the most relevant roughness parameter that characterizes the surface morphology of a manufactured product with regard to a correlation with a particular function, property or application [1]. Moreover one other important challenge we have to take up is to define the scale at which one parameter must be evaluated to be pertinent. This led us to observe and analyze the problems of existing software:

1. Are not adapted to deal with a data file.
2. Only a limited number of parameters.
3. Are unsuitable for statistical analysis.
4. Often a system designed for a specific measure.
5. Requires too many manipulations.
6. Reliability of certain procedures?
7. No non-linear algorithm.
8. Sustainability software.
9. Do not possess multiscale treatments.

These many questions have no answers at present. Indeed, software trade can not offer a solution to the raised issues. For these reasons, we decided to design our own surface status analysis system called "MesRug"

2. Presentation of MesRug

This environment for surface conditions analysis (so-called "MesRug") is a software processing roughness files. The analysis starts with the experimental design construction. Each process parameter set is in relation with a measurement data file. The save of time in processing with MesRug is significant. We illustrate this by a simple example. Profile measurements are carried out on the sample. The purpose of this analysis is to answer to the question: what are the roughness parameters that best discriminate process parameters and at which scale must they be evaluated? We enter all necessary information in "MesRug" and launch calculation of roughness parameters. "MesRug" processes all files, creates text files for further subsequent analysis (spectrum, autocorrelation function…) and graphics files while preserving the coding performed by the operator. The result files are encoded to be used in Excel, Statistica and SAS (Statistical Analysis System). The statistical processing is done under the SAS language.

3. Structure and Construction

This system is built from a database which allows describing, manipulating, processing the data sets and can be used by non-scientists. Fig. 1 diagrams the links between the various parts of the data treatment.

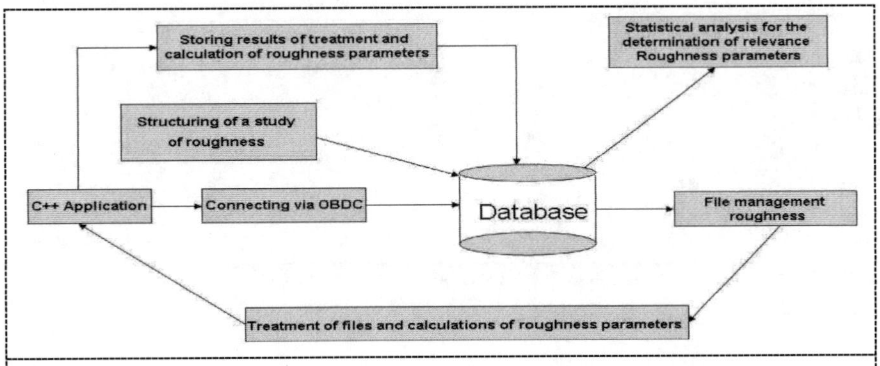

Figure 1. Structure of the MesRug software.

3.1. Recovery profile and calculation of parameters

The database is programmed in C++. Open DataBase Connectivity (ODBC) enables to treat data files via Access, ActiveX Data Object (ADO) and SQL. We build a structure that allows to realize a multiscale pretreatment (polynomial interpolation, Gaussian filter, Bspline, wavelet transform…) to transform initial profile in order to calculate roughness parameters. This structure is constructed in an open source concept to allow in the future an easy integration of new procedures and pretreatment process.

3.2. The script language

The basic idea we developed is to use a script language that allows to detail the structure of the analyses treatment . Then all computations are applied to the set of roughness file without manipulation. To organize this structure, a recursive scheme of data is proposed. Rather to generalize our data structure, we will treat a simple case of profile. For a given profile, we process to the following treatment: first a low Gaussian pass filter and a high pass Gaussian filter are applied on the initial profile, then a multiscale analysis is processed on these filtered profiles with two characteristic lengths (10 and 100 intervals). As a consequence, one gets 1x2x2=profiles on which roughness parameters can be computed. Then a R_a and R_q and fractal dimension (by DF method (structure method) with some scale parameters) are computed on these profiles.

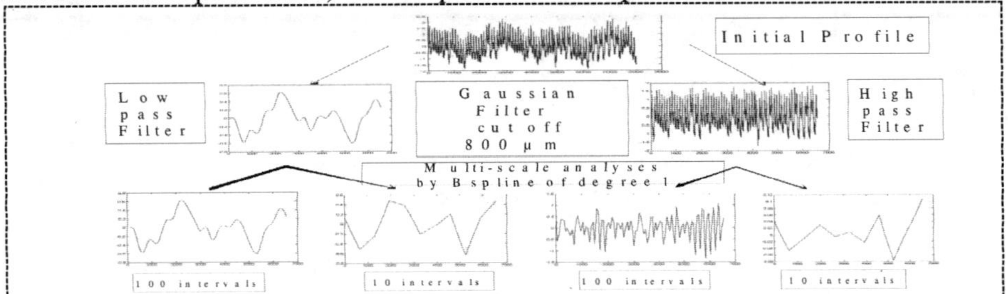

Figure 2. Multiscale analyses on a tooled surface.

The recursive scheme can be formulated as follows:

Possible de mettre la formule en plus grand?

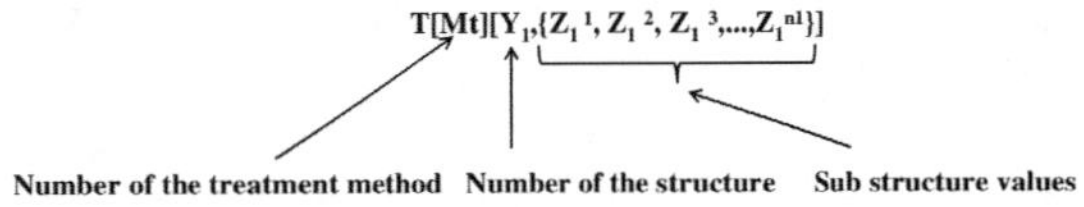

In our case, one gets the following script :
// Gaussian Filtering (Vorburger algorythm)
T[3][1,{800}]
T[3][2,{passe_haut,passe_bas}]
// Bsplines analyses
T[2][1,{1}]
T[2][2,{10,100}]
// R_a, R_q and fractal dimension using 100 scales from 10 to 10000 R[Ra]
R[Rq] // windows following a geometrical progression ({2}).
R[DF,{10},{10000},{1000},{2}]

And the result is given is our database (table 1)

Table 1. Result of multi-scale analyses stored in our DataBase thanks to ODBC protocol

File	Filter	Scale	R_a	R_q	DF_10_10000_1000_2
1_A_C_1.std	Passe_haut	10	0.542	0.658	1.542
1_A_C_1.std	Passe_haut	100	0.295	0.372	1.596
1_A_C_1.std	Passe_bas	10	0.057	0.070	1.070
1_A_C_1.std	Passe_bas	100	0.001	0.001	1.027
1_A_C_2.std	passe_haut	10	0.533	0.647	1.532
....	....	....	....	....	....

4. Evaluation of the relevance of the roughness parameters and their associated range

As the database is created with all values of multi-scale aspects, the main purpose is to analyze what are the best roughness parameters to describe physical phenomena and their associated scales. In fact, the answer to this question is not so obvious and must be formulated in a statistical sense. Several cases are presented and treated with appropriate statistical tools. In this paper, only two cases will be treated.

An experimental design is built to create some surfaces (as tooling ....) and the question is :

What are the influence of the process parameters on surface topography ?

This question is reformulated as follow in surface topography formalism:

What are the best roughness parameters evaluated on an appropriate scale that allow to better discriminate the effect of the process on the surface topography ?

This question must be also reformulated in a statistical sense. In this case, the analyses of variance will be used. This method allows giving the influence of each ith process parameter p_i (with $i \in \{1.\cdots I\}$) on the value of the jth roughness parameters q_j (with $j \in \{1.\cdots J\}$) evaluated at the kth scale ε_k ($k \in \{1.\cdots K\}$) that is noticed $q_j(\varepsilon_k)$. To test the relevance of $q_j(\varepsilon_k)$ to well characterize the effect of the I process parameters p_i, $K \times J$ analysis of variance are performed for all J roughness parameters and all K discretized scales. The Fisher variate F is used to quantify the effect i.e. the relevancy and is noted $F_{p_j}(q_j(\varepsilon_k))$. If $F_{p_j}(q_j(\varepsilon_k)) > 1$ then $q_j(\varepsilon_k)$ roughness parameters is relevant to characterize the p_j process parameter. The higher $F_{p_j}\left(q_j(\varepsilon_k)\right) > 1$ is, the more relevant $q_j(\varepsilon_k)$ roughness parameter is. It can be specified that the interaction between process parameters q_i and $q_{i'}$ can also be computed and noticed $F_{p_j,p_j'}\left(q_j(\varepsilon_k)\right)$.

F value allows us to gather the roughness parameters that possess the same relevance on a range scale $[\varepsilon_k, \varepsilon_{k'}]$ rather on a unique one. Let illustrate now our purpose by one example based on belt finishing process taking account 7 process parameters (wheel stiffness, pressure, oscillation, grit size…) [2]. For the wheel stiffness, the best roughness parameter is the S_m (distance between asperities). Let now analyze the results by plotting the $F_{stiffness}\left(Sm(\varepsilon_k)\right)$ versus ε_k (figure Fig. 3 on the left). This curve presents 3 local maxima. One of the extreme is based on the very high scale (around 2500 µm). At this scale, S_m is higher for the softer wheel (result not shown). This tendency is due to the elastic deformation of the wheel during contact that creates some waves of low frequency on roughness of materials. At the scale of 7.5 µm (microscopic scale), S_m is higher for the softer wheel. At this scale, roughness is analysed on a scale less that the mean size of the grooves and then the seen roughness is the groove. Into the groove, ploughing wear is higher with soft wheel (see fig. 3) and naturally increases the mean size between asperities. On the contrary, at the scale of 30 µm, S_m is higher for the higher stiffness. At this scale, roughness is analysed on a scale higher than the mean size of the grooves and then characterizes the groove's width. When stiffness wheel increases, indentation depth increases and S_m naturally increases.

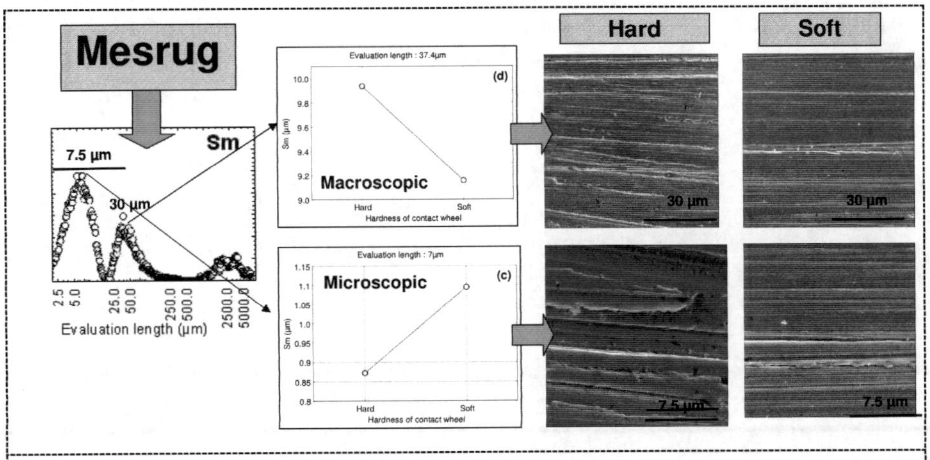

Figure 3. Analyses of the influence of wheel stiffness on S_m roughness parameters.

The second example is linked to the functionality of the surfaces and can be also treated by our software. Let answer to the question: what is the best roughness parameter that characterizes the wettability of titanium surfaces? This problem is reformulated in a statistical sense "what is the best relation between a roughness parameter evaluated at a given scale and the measure of wettability noted w i.e. $w = f\left(q_j(\varepsilon_k)\right)$ where f is a given functional (as linear one). In this case, we find the best correlation by a linear one characterized by a Student variate (P value). With this result, it was found that the best parameter to characterize contact angle is R_v (valley amplitude) and that the functional *f* is linear and evaluated at 400 µm (fig. 4). This result was explained by the Cassie Baxter theory rather by the Wenzel one.

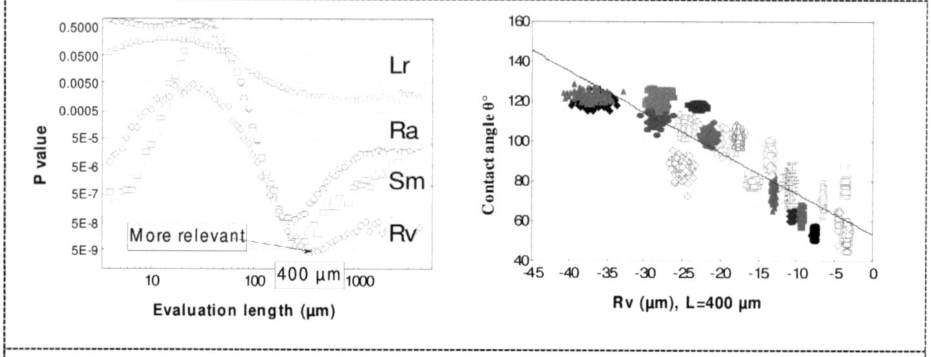

Figure 4. Multi scale Mesrug analyses (left) of the wettability on titanium surface roughness (right).

5. Conclusion

We have constructed a methodology to select roughness parameters with regard of surface functionalities. It was shown that relation between parameters can depend on the scale. Thanks to our expert system Mesrug, it was possible to quantify the different scales at which parameters must be evaluated. Applied on different surface engineering studies, it was then possible to explain how surface topography influences the integrity (abrasion) or functionality (wettability) of materials. A 3D version of Mesrug is under construction.

6. References

[1] Najjar D, Bigerelle M and Iost A 2003 *Wear* **254** 450

[2] Van Gorp A, Bigerelle M, El Mansori M, Ghidossi P and Iost A 2010, *Int. J. Mater. Prod. Tech.* **38** 66

Acknowledgments

This work was realized within the framework of the project "Functional Impacts of the three-dimensional rough surfaces" financed by the region "Picardie" and Europe. We want to thank Gildas Guillemot and Adrien Van Gorp for their contributions to the programming of Mesrug Software.

3D finite element model of elastoplastic contact on the double sinus rough surface

H HAGEGE[1], S BOUVIER[1], P-E MAZERAN[1] and M BIGERELLE[1]

[1] Laboratoire Roberval, UMR 6253, UTC/CNRS, UTC Centre de Recherches de Royallieu BP 20529, 60205 Compiègne, France.
E-mail: benjamin.hagege@utc.fr

Abstract. One of the objectives in the field of tribology is to solve the mechanical stress-displacement problem involved by rough contacts. In our approach, the surface chosen is a 256-256 µm² 3D sinusoidal shape (amplitude 4.5µm, wavelength 50µm) with an elastoplastic constitutive behaviour. The constitutive law combines isotropic and kinematic hardening and is experimentally identified from 316L steel sheets. The FEM deformable surface is crushed then uncrushed by a rigid flat surface: stresses, contact pressure and plastic cumulated strain are computed. We investigate the results sensitivity with respect to the level of in-plane refinement. At last, we conclude on some guidelines for 3D finite elements modelling of rough surfaces.

1. Introduction

Tribology analysis of surfaces is traditionally associated with stochastic and semi-analytical models. But the assumptions made are sometimes too restrictive because of 3D geometry of surfaces and non-linear elastoplastic behavior of materials. Nevertheless, in order to model more precisely the response of a rough surface, many authors modeled the contact with Finite Elements Method (FEM). However, because of the large size of such numerical models, computations are still often restrained to 2D plane-strains slices, from which the surface is supposed to be 'extruded', despite 3D approaches have been already tried by [1]. 3D models allow to estimate with more accuracy the stress field and to take into account the "geometric coupling" between asperities, but computation times dramatically increase with mesh density. However, by means of new shared memory computing blades and with adequate softwares (i.e. Ansys Academic Research™), it becomes possible to analyse 3D elastoplastic contact of realistic discretized surfaces. Our objective is to develop a finite elements mechanical computation methodology of rough surfaces.

2. Materials and the finite element model

The studied surface is showed on Fig. 1: it is simple because of its geometry regularity and it is a reference case because of the sinus asperities shape [2]. Because of the "low" roughness of this surface, the finite elements mesh is obtained by perturbation of a pre-existing mapped mesh. Perturbation consists in the displacement of the surface nodes by spatial detection and comparison to an input file coming from a profilometer device. The low R_t parameter (compared to the wavelength of an asperity) allows the generation without "folded" elements.

1 To whom any correspondence should be addressed.

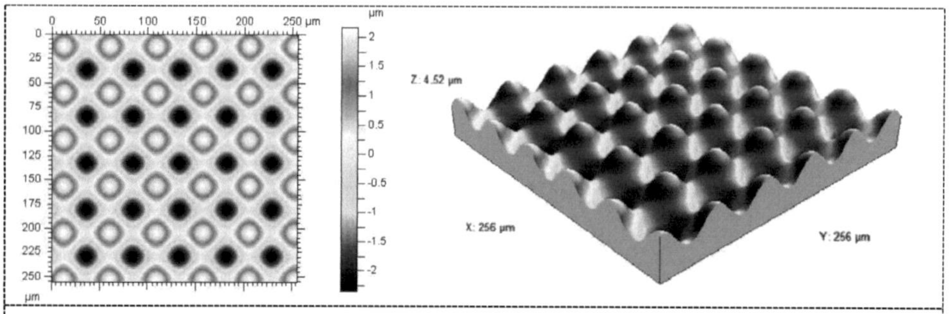

Figure 1. Home-made double sinus surface

2.1. Constitutive behavior

The material is a 316L steel with an initial yield stress of 212.54MPa described by the Lemaître-Chaboche hardening: isotropic Voce hardening (saturation Rinf=1000MPa and exponent b=1.6) plus one term Armstrong-Frederick kinematic hardening (parameters C1=4836.4MPa and G1=23.5). Young modulus is 200GPa and ν=0.3. The 1D material law in compression then tension is showed on Fig. 2.

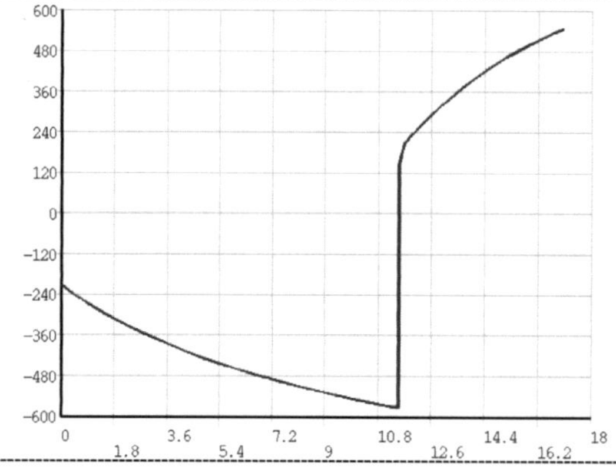

Figure 2. 316L constitutive behavior in the Von Misès (MPa) – accumulated plastic strain (%) plot

2.2. Finite elements meshes and boundary conditions

The input file is done with 1µm space increments in X and Y directions. As the surface is a 256µm square, maximum density corresponds to a mesh size of 256 x 256 elements on surface, like mesh 3, whereas meshes 2 and 1 are successively twice less dense (see Fig. 3). Mesh 101 corresponds to mesh 1, but with parabolised H20 elements coming from the H8 mesh of 1. All these meshes have a thickness of 256/8.5µm=30.12µm.
The modeled substrate comes from a sample (dimensions: 1cm²x1cm) that is free on its four sides and that lies on a plane on its underside. Because of symmetries, we also take free edges for the substrate, however, the substrate bottom is simply supported in the Z direction, leading in overestimated stiffness, and so are the stresses and plastic strains. The rigid surface crusher is displacement driven

along Z with all other degrees of freedom blocked. Surface goes down to $R_t/4$=1.13μm and goes back up to 0.904*$R_t/4$ to finely analyse the spring-back. Regarding the interface behavior, it is desirable to employ an exact contact algorithm since the problem is dominated by contact mechanics. This is why the Pure Lagrangian algorithm is chosen. There is no friction.

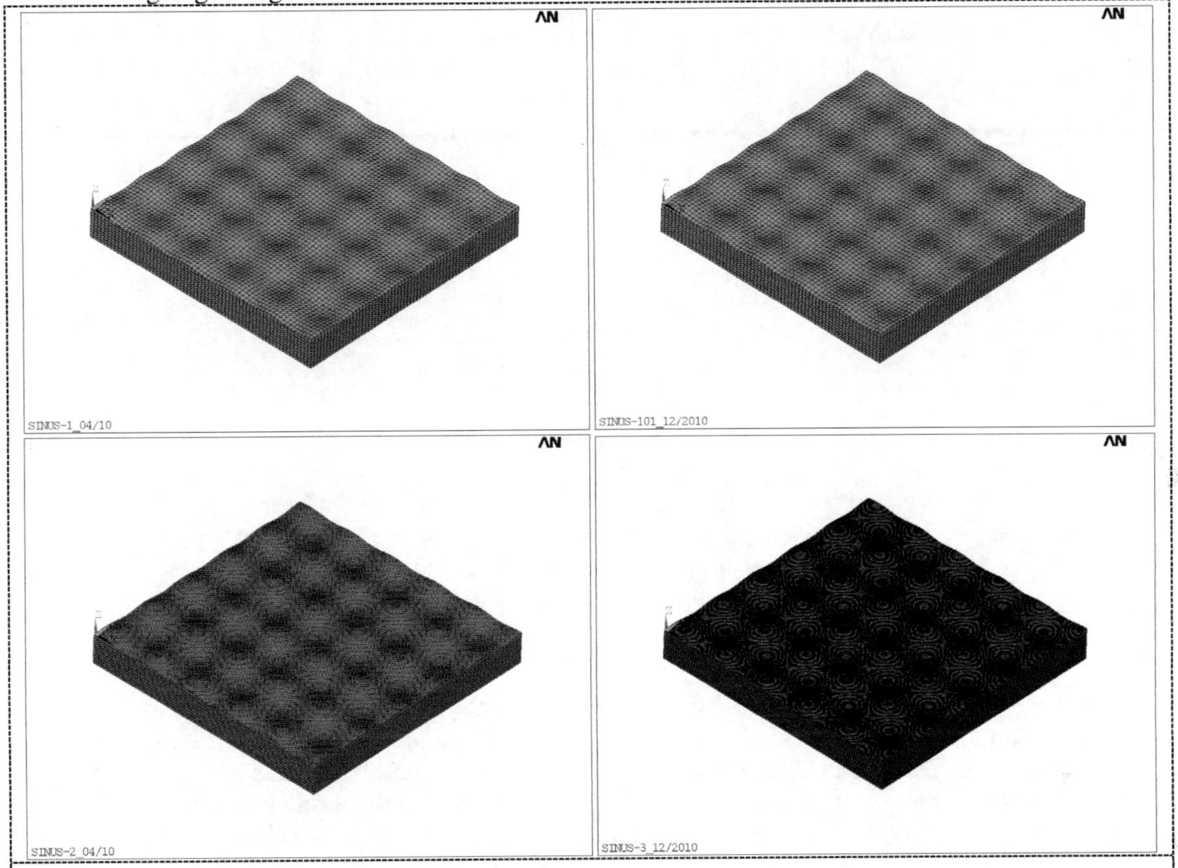

Figure 3. Meshes 1 to 101 and 2 to 3, from top to bottom, from left to right

2.3. Model parameters and numerical performance

We summarize the different models on Tab. 1 and we emphasize the large numerical resources needed to run them. The supercomputer used was a BULL 6030 Supernode with 24 X7542 Xeon cores, 256Gb RAM and 3.2To stripped 10kRPM HD. All models have 10 elements in the Z thickness direction.

Table 1. Parameters and numerical performances

JOB	MESH		NUMERICAL PERFORMANCES			
Nb	Nb elem. X	Nb elem. Y	Nb dof	Nb Core	Comp. Time (s)	Required RAM to run incore (Mo)
1	64	64	132345	5	2732	945
101	64	64	519919	4	42104	8017
2	128	128	526841	6	15794	4759
3	256	256	2096612	6	100604	23947

3. Convergence analyses

3.1. Convergence of macroscopic out of plane force and contact area

Fig. 4 highlights a quick convergence of the FZ reaction force of the rigid surface and of the total contacted area: model 2 with 128x128 elements on surface is suitable whereas models 1 and 101 with 64x64 elements are too coarse.

Despite it is not needed, model 3 is the best but at a large computing cost (see Tab. 1).

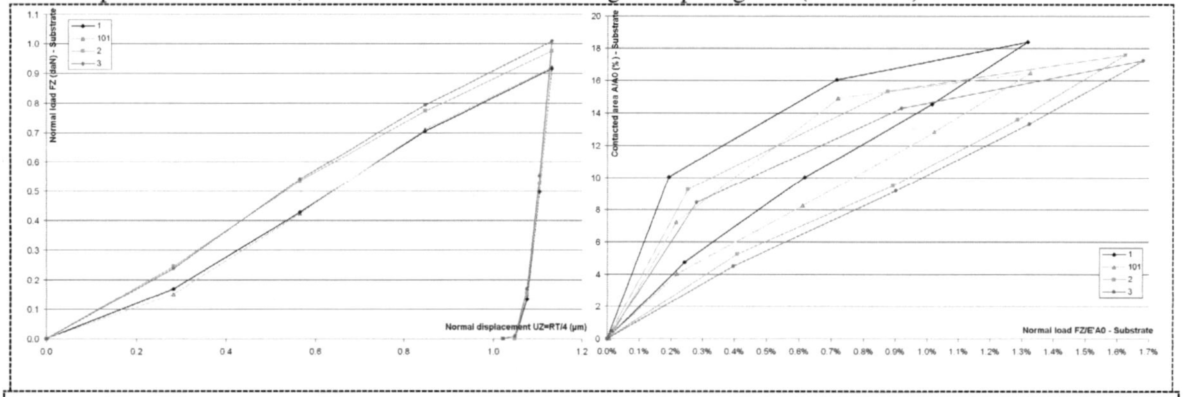

Figure 4. Force-displacement curves of the rigid surface (left) and contacted area curves of the whole deformable surface (right)

3.2. Convergence of the Von Mises stress at the end of the crush step

On one hand, we show on Fig. 5 that only model 3 allows to obtain the fine distribution of in plane surface maximum values that are the crowns around the maximum contact areas (between 400MPa and 450MPa in orange range). On the other hand, the maximum Von Mises stress is found beneath the surface, below the top of asperities as expected by extending the classic elastic Hertz theory of contact to elastoplasticity. All meshes localize well this stress peak and lead to values between 470 and 491MPa.

Model 3 is the best and is needed for good in plane discretization.

4. Conclusion

We have demonstrated the feasibility of 3D finite elements computations in the case of the regular double sinus rough surface. We may conclude on guidelines to start to build a general methodology. Here are the 2 main conclusions:

-quadratic elements are not recommended because they do not improve the discretized contact state, that must be finely described, while they dramatically increase computing time,

-in order to finely discretize the in plane Von Mises stress field, fine meshes with in plane 256x256 element densities are needed (element size = 1µm that is 50 elements / wavelength here).

Thereafter, it will be necessary to analyze residual stresses at the end of the "uncrush" step. Moreover, to have a more realistic 3D approach, friction and sample stiffness must be taken into account with the aid of a spring.

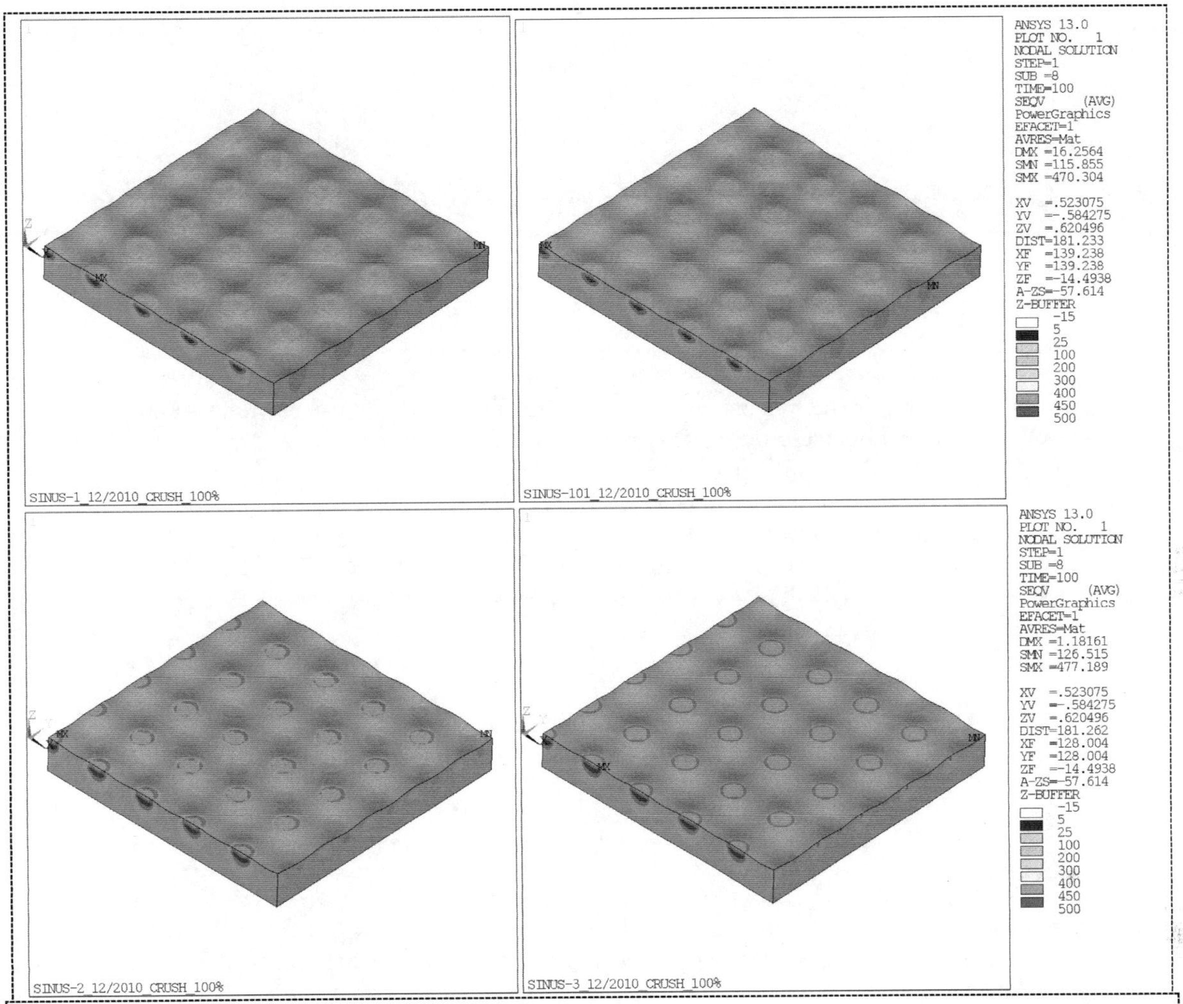

Figure 5. Von Mises stress field for meshes 1 to 101 and 2 to 3, from top to bottom, from left to right

5. References

[1] Mahrenholtz O, Bontcheva N and Iankov R 2005 *J. Mater. Proc. Tech.* **159** 9
[2] Söderberg A and Björklund S *Trib. Int.* **41** 926

Acknowledgments

The authors acknowledge the Projet Pluri-Formations PILCAM2 at the Université de Technologie de Compiègne (UTC) for providing HPC resources that have contributed to the research results reported within this paper. URL: http://pilcam2.wikispaces.com.

How to characterize the regularity of surface topographies?

G GUILLEMOT[1], M BIGERELLE[2], J ANTONI[2] and K ANSELME[3]

[1]Laboratoire de Mécanique de Lille, UMR CNRS 8107, Arts et Métiers ParisTech - Lille, 8, boulevard Louis XIV 59046 Lille Cedex, France.
E-mail: gildas.guillemot@ensam.eu
[2] Laboratoire Roberval, UMR 6253, UTC/CNRS, UTC Centre de Recherches de Royallieu BP 20529, 60205 Compiègne, France.
[3] Institut De Sciences Des Matériaux De Mulhouse, CNRS LRC 7228, 15, rue Jean Starcky, Université De Haute-Alsace, BP 2488, 68057 Mulhouse, France

Abstract. In this paper, a new parameter is proposed to characterize the regularity of a surface topography. The regularity of the surface with scaling invariance both in amplitude and position allows us to quantify the organization of surfaces by a number lying between 0 and 100%. This parameter is used to quantify the influence of different processes on the topography of the surface and calibrated on noisy periodical surface topography.

1. Introduction
Current literature does not provide su any parameter to quantify the regularity of the surface with scaling invariance both in amplitude and position. Consequently, we created a new original scale invariant roughness parameter to quantify the organization of surfaces. This parameter is used to quantify the influence of different processes on the topography of the surface.

2. The regularity parameter
The main idea is to find a parameter without resorting to Fourier's analysis because spectrum parameters have little robustness and sin-cosine basis is not always appropriate to characterize surface roughness. This parameter must have a upper limit value (100%) if surfaces are periodic, a medium value if surfaces get a non neglected first order autocorrelation and lower limit null value for uncorrelated random surfaces (white noise). First, we define a normalized autocorrelation function such that:

$$R(x_i) = \frac{1}{Rq^2(N-i)} \sum_{j=1}^{N-i} y_j y_{j+i} \quad (1)$$

where $(x_i, y_i)_{,\ 1 \le i \le N}$ are the N equidistant discretized points of the roughness profile and Rq is the standard deviation of the amplitude. The autocorrelation length, L, corresponds to the first point x_i ($L = x_i$), such that $R(x_i) \le \lambda$, where λ is a chosen threshold. A threshold value of 0.1 is usually chosen. Secondly, we define a correlation integral J such that:

[1] To whom any correspondence should be addressed.

$$J = \int_{x=0}^{x=L} R(x)\,dx \tag{2}$$

This integral represents a kind of fundamental for the autocorrelation function.
Thirdly, we define the series of integrals I_K :

$$I_K = \int_{x=k\,L}^{x=(k+1)\,L} |R(x)|\,dx \tag{3}$$

These integrals represent the successive harmonics. Finally, the regularity parameter is defined as:

$$Regularity = \frac{100}{KJ}\sum_{k=1}^{K} I_k \tag{4}$$

This parameter lies between 0 (white noise profiles) and 100 (perfect periodic profiles without noise). The main advantage of this parameter is to be mathematically independent of the amplitude parameter and of the autocorrelation length, L, of the surface. This parameter could be extended in two dimensions and used to measure the anisotropy of surfaces. We intensively used this parameter in profile analysis. Moreover it can be linked to the information of surfaces via Shannon's theory.
To test the relevance of this parameter to measure regularity, some simulated profiles have been defined. These profiles are noised with a stationary uncorrelated Gaussian noise (*ie. white Gaussian noise*) with a given standard deviation. Various shapes of periodical function have been retained: a sinusoidal function ('*Sinus*'), a triangular function ('*Triangular*'), a square function ('*Square*'), an shifted absolute sinusoidal function ('*Abs Sinus*'), and finally A white noise (see figure 1, left). Profiles have 10000 points and are normalized in the range [-1;1]. Then the effect of the added Gaussian noise is analyzed (see figure 1, right). 100 profiles are simulated for each calculation.

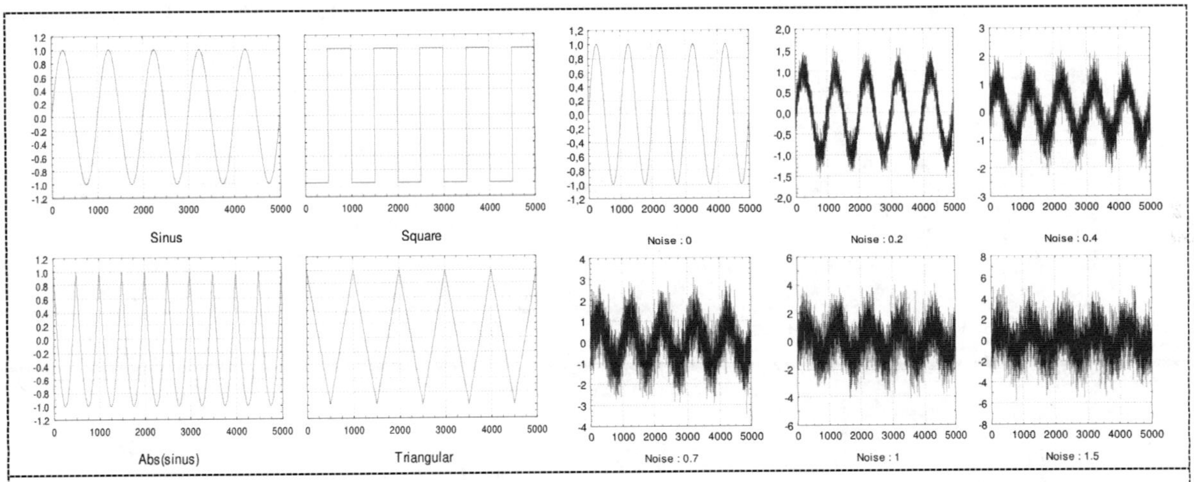

Figure 1. Periodical Functions used to test the Regularity parameter. The functions have been represented in the range [0…5000] (left) and Sinus Functions test with added white Gaussian noise at various amplitude levels. The 'Noise' value corresponds to the standard deviation of the added noise (right).

Figure 2 represents the **A**utocorrelation **F**unction (AF) of the sinus function for different values of the noise (Fig. 1). The AF of the unnoised sinusoid is the sinusoid function itself. The amplitude of the AF for x>0 decreases by a constant factor when the amplitude of the added Gaussian noise increases. Nevertheless AF(x=0) is always equal to 1.
To compute the regularity parameters, the **A**utocorrelation **L**ength (AL) must be defined. The definition of the autocorrelation length is still under debate. Its value is often affected to the fact that the value of the lag length at which the AF drop to 1/e of its value at zero lag and will be called the

Inverse **L**ag **L**ength λ (**IAL**). This value depends of the threshold value used to equal the autocorrelation function. This threshold value will further be called the **I**nverse **L**ag **L**ength (**IAL**), λ. Sometimes, authors take the IAL as equalled to 0.1 or 0.4. The only justification to the value of 1/e (ie. 1/2.71828~0.4) is that AF follows an exponential form and then analytical computation will be simplified in a high number of scientific fields. The value of the regularity parameters depends on the autocorrelation length and it will be of major interest to compute the value of this parameter for various AL and corresponding IAL threshold values.

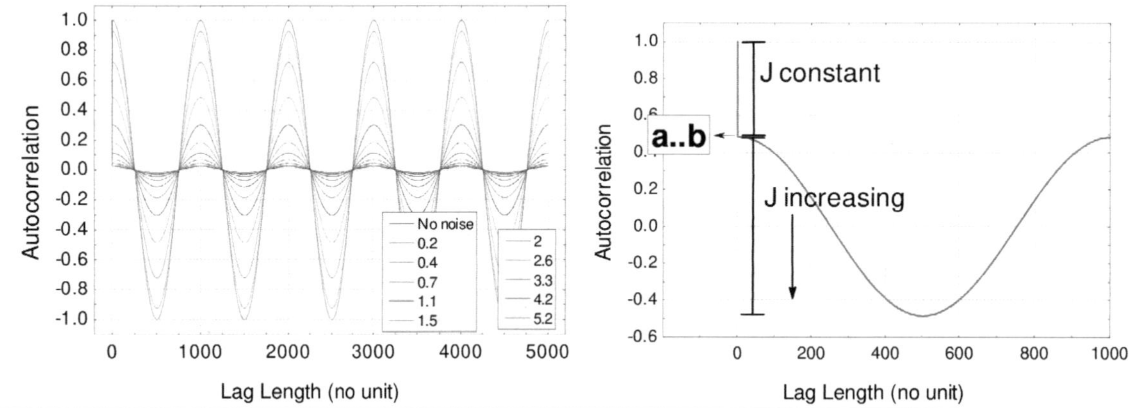

Figure 2. Autocorrelation function of sinus functions test with white noise at different amplitude level (left) and an zoom of Autocorrelation function of a noisy sinus function (right) with notations used in the text.

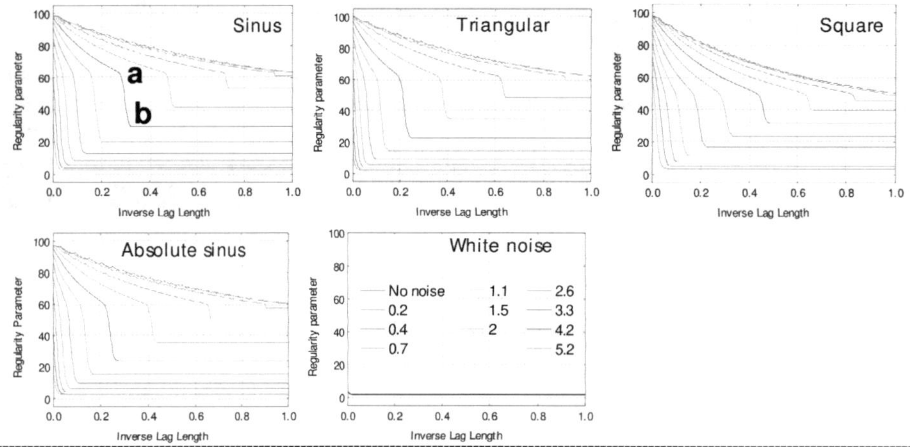

Figure 4. Value of the regularity parameters versus the inverse lag length for the six functions with different Gaussian noise amplitudes

Figure 3 represents the value of the regularity parameters versus the λ that determines the autocorrelation length allowing the computation of the regularity parameter for the six functions with various Gaussian noise amplitudes (Fig. 1). Let now analyze the various results:
Regularity functions always decrease or stay constant for a given noise when λ increases.
For all unnoised periodical functions, for $\lambda = 0$, the regularity is equals to 100%.
For noised periodical functions, it can be observed three stages, $\lambda \in [0,a]$, $\lambda \in [a,b]$ and $\lambda \in [b,1]$:
Stage 1, $\lambda \in [0,a]$: A decrease until reaching $\lambda = a$. For $\lambda = 0$, the regularity decreases with an increase of noise amplitude until reaches the value of the noise regularity (regularity = 1.9 +/- 0.001). The regularity is maximal for $\lambda = 0$, because the integration of J and I_K is computed on the ¼ period of the functions. For $\lambda > 0$, the regularity decreases with an increase of λ. This decrease is due to the fact that taking a value of $\lambda > 0$ will decrease the I_K value.
Stage 2, $\lambda \in [a,b]$: The a value depends on the noise amplitude and decreases when noise amplitude increases. The value of regularity (a) is a noise-amplitude independent for a given function and is equals to regularity (1) of the unnoised function. This fact can be explained as follow: while $\lambda > b$, the J integral is constant and measures the noise (figure 2, left). When $\lambda < a$, then J is integrated on the part of period of the autocorrelation function increasing then the regularity. The interval [a,b] is due to stochastic sampling by increasing the number of points in the profile, the difference a-b decreases.
Stage 3, $\lambda \in [b,1]$: Regularity function is constant because J is always constant and I_K. As I_K decreases when noise increases, regularity function decreases.

We have explained why regularity function depends on the choice of the Inverse Autocorrelation Length (IAF) value, λ. However, the choice of the value is still under debate. As consequence, the relevance of the choice must be shown. It is stated that the best choice of λ value will enable the best discrimination in change of the profile. In our case, the change of profile regularity is given by the noise amplitude and then we have to find the λ value which will give the best discrimination between the various noises. Consequently, an analysis of variance will be performed on the regularity parameter at each λ value for all classes of noise amplitudes. The Fisher statistics $F(\lambda)$ is then computed. The greater $F(\lambda)$ is, the better regularity parameters discriminate noise effect on the profiles. Figure 4, (left) represents the F values for the five periodical signals, versus the IAF λ. Some results are shown on these curves:

- The relevance depends on the IAF and consequently this value can not be randomly chosen.

- The relevance is very poor for the low IAF value (typically λ<0.1). This result is quite surprising because for IAF equals to zero, the integrations are computed on ¼ of the period and would decrease the variance of the estimator. In fact, the variance of the auto correlation length is maximal and increase the variance of the regularity parameter.

- Even if curves seem similar, regularity better discriminate the sinus, triangular and absolute sinus than square and triangular one. The both last periodical function presents the well known Gibbs phenomena (discontinuity) that increases the variance of the estimator of the regularity.

- When AIF is estimated near the unity, the regularity parameter well discriminates the noise influence.

- The curves possesses some local decrease of relevance: at some particular values of the AIF, due to the fact that for some particular noise, the autocorrelation length can be change from small value (noise) to a ¼ period (function) that increases variance of the regularity estimation.

Journal of Physics: Conference Series **311** (2011) 012012 doi:10.1088/1742-6596/311/1/012012

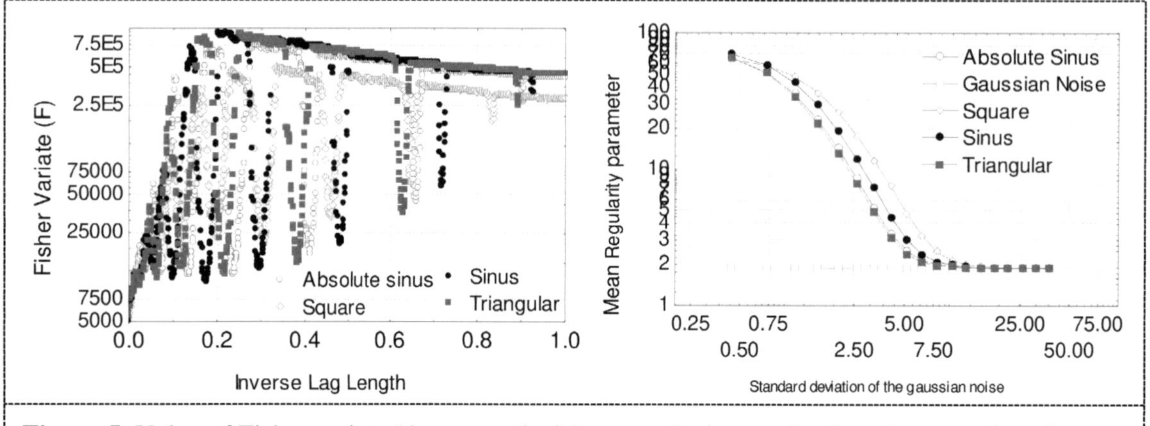

Figure 5. Value of Fisher variate (Anova analysis) versus the inverse lag length to test the relevance of the regularity parameter to discriminate the effect of noise on test functions (left). Autocorrelation function of sinus functions test with white noise at different amplitude level(right).

As a consequence, due to the difficulty to find accuracy the autocorrelation length, we propose to avoid the use of a single IAF and a correspondingly single autocorrelation length. We then propose to retain the average of the regularity parameter computed for all autocorrelation length (given at different level from 0 to 1 of the autocorrelation function). Then the new Mean Regularity roughness parameter $\overline{O}$ is proposed:

$$\overline{O} = \int_{0}^{1} 100 \sum_{k=1}^{K} \int_{x=kL(y)}^{x=(k+1)L(y)} \left|R(x)\right| dx \Bigg/ \left(K \int_{x=0}^{x=L(y)} R(x)\, dx \right) dy \qquad (5)$$

with $L(y)$ is defined such that $y = R(L(y))$ is the autocorrelation length taken at the value y of the autocorrelation function. Figure 4 (right) represents the evolution of the mean regularity parameter $\overline{O}$ versus the intensity of the Gaussian noise. This parameter discriminates for all periodical functions the effect of the noise: higher is the noise intensity, lower is the regularity of profiles. To test the accuracy, the analysis of variance is computed to quantify how this new parameter discriminate the effect of the noise. Table 1 represents the F values for the five periodical functions. As it can be observed, Values of the F is quiet less than the maximal value corresponding to the most relevant correlation length but is better in mean than choosing randomly the Autocorrelation length. This step validates the relevancy of the mean regularity parameter applied on periodical function.

Table 1. Value of Fisher variate (Anova analysis) for the mean regularity parameter to test the relevance to discriminate the effect of noise on test functions. The max F and Mean F respectively represent the maximal values obtained from the figure 5.

	Absolute sinus	Sinus	Triangular	Triangular one	Square
$\overline{O}$	$6.54\ 10^5$	$6.78\ 10^5$	$6.64\ 10^5$	$4.73\ 10^5$	$4.88\ 10^5$
Max F	$7.58\ 10^5$	$9.21\ 10^5$	$9.11\ 10^5$	$3.44\ 10^5$	$4.99\ 10^5$
Mean F	$4.01\ 10^5$	$4.39\ 10^5$	$9.11\ 10^5$	$4.41\ 10^5$	$2.63\ 10^5$

This methodology was applied on various surface topographies and demonstrates that the most relevant roughness parameters are frequencies ones rather than amplitude ones to characterize cell adhesion. It is also proves that periodicity of the surface decrease cell adhesion [1].

3. Conclusion

By this new parameter used to characterize the morphological properties of a surface, it was shown that the effect of quality of the periodicity a profile can be determined by an unscale parameter. We works actually on weighting the autocorrelation function according to its variance to introduce a corrected Regularity parameter that will certainly diminish its standard error. Other aspects are still under discussions such analyzing the scale effect on fractal curves and the effect of autocorrelated noise on periodical signals.

4. References

[1] Anselme K, and Bigerelle M, 2005 *Acta Biomaterialia* **1(2)** 211

Acknowledgments

This work was realized within the framework of the project "Functional Impacts of the three-dimensional rough surfaces" financed by the region "Picardie" and Europe.

Journal of Physics: Conference Series **311** (2011) 012013 doi:10.1088/1742-6596/311/1/012013

Wavelet theory and belt finishing process, influence of wavelet shape on the surface roughness parameter values

Z khawaja[1*], G Guillemot[2], P.-E Mazeran[1], M El Mansori[3], M Bigerelle[1]

1 Laboratoire Roberval, UMR 6253, UTC/CNRS, Centre de Recherche de Royallieu, 60205 Compiègne, France

2 Laboratoire Mécanique de Lille, UMR CNRS 8107, Equipe Freinage Contact Surface, Arts et Métiers Paris Tech, 8 Bd Louis XIV, 59046 Lille Cedex, France.

3 Laboratoire de Mécanique et Procédés de Fabrication (LMPF-JE2381), Arts et Métiers Paris Tech, Rue Saint Dominique, BP 508, 51006, Châlons en Champagne, France

Zahra.khawaja@utc.fr

Abstract. This article presents a multi-scale theory based on wavelet decomposition to characterize the evolution of roughness in relation with a finishing process or an observed surface property. To verify this approach in production conditions, analyses were developed for the finishing process of the hardened steel by abrasive belts. These conditions are described by seven parameters considered in the Tagushi experimental design. The main objective of this work is to identify the most relevant roughness parameter and characteristic length allowing to assess the influence of finishing process, and to test the relevance of the measurement scale. Results show that wavelet approach allows finding this scale.

1. Introduction

Precision tooling by turning and rectification of functional surfaces of mechanical parts, i.e. obtaining flawless parts both geometrical and structural, requires a considerable technical and economical effort. Some operations are completed by a finishing process such as the belt grinding process which is easier and less expensive than others finishing processes.

The difficulty is the incomplete knowledge of both the mechanisms and the characteristics of this process. Some authors analyzed the relations between the conditions of this process and the roughness of the surfaces obtained by experimental approaches [1, 2].

2. Experimental study

This process consists in applying an abrasive oscillating belt, of low thickness, on a rotating manufactured specimen. To assure the reproducibility of the process, five bearings, having a diameter of 54.78 mm and a width of 30 mm, are tooled for each group of test conditions of the experimental design. The width of the belt is 20 mm. Tagushi's experimental design is used to study the effects of the process conditions on the resulting roughness. The experiment was conducted using 16 work specimens that were turned and rectified. All samples are then manufactured with the same lubrication condition corresponding to '*Cut Max H05*'. We retain 7 process parameters that are Belt feed (50 and 100 mm/mn), contact pressure (1 and 3 bars), Axial oscillation frequency (1.6 and 10 Hz), Contact wheel stiffness (hard and soft), Cycles times (3 and 9 s), Belt Grit Size (9 and 40 µm) and finally Work piece rotation speed (100 and 500 rpm) [3].

This experimental design is built in a way to consider the contact wheel hardness versus the contact pressure, the contact wheel hardness versus the belt grit size, and the belt grit size versus the contact pressure. Roughness measurements are performed for each specimen. Thus, 30 roughness profiles were recorded from the tooled surfaces by a KLA-TENCORTM P-10 profilometer with a 2 µm tip radius. The scanning and the sampling lengths are respectively 8 mm and 0.1 µm.

3. Wavelets transform

Wavelet analysis is a mathematical tool recently developed for the signal treatment [4]. The main idea behind this analysis is using the wavelet functions satisfying certain mathematical requirements to analyze signals [5]. This approach uses windows with short width for the high frequencies and windows with large width for the low frequencies. Thus a multi-scale spatial analysis is developed.

The procedure of wavelet analysis begins by choosing a prototype wavelet function, called mother wavelet Ψ, to produce the other functions of different windows. This basic function will be translated and dilated to cover the plan time-frequency and analyze the signal.

By introducing the factors of translation, b, and of scale, a, we obtain the wavelet girl, $\Psi_{b,a}$:

$$\Psi_{b,a}(X)=\frac{1}{\sqrt{a}}\Psi\left(\frac{X-b}{a}\right) \quad (1)$$

So due to its properties of dilation-contraction and translation, the wavelet transform (WT) is characterized in the space-scale plane by a window of variable width. This width decreases when we focus on the structures of small scale (high frequency) or increases when we are interested in the large-scale behaviour (low frequency) [6].

Wavelet transforms are broadly divided into three classes: continuous, discrete and multiresolution-based. In our implementation the discrete wavelet transform (DWT) is used. This wavelet transforms defined by the following equation:

$$W_{i,j}(X)=2^{j/2}\Psi\left(2^{j}X-i\right) \quad (2)$$

Where i and j are integers and $W_{i,j}(X)$ is the wavelet coefficient calculated in the scale j and at the position i. The DWT is based on a filtering algorithm developed by S. Mallat [7] in 1989. He considered the wavelet analysis as a decomposition of the profile by a cascade of filters, associating a pair of filters with every level of resolution. The profile is decomposed as a consequence into an approximation, A_i, and details, D_i, information. The approximation is then decomposed using the same wavelet decomposition. This is achieved by successive high pass and low pass filtering of the time domain signal. This algorithm of decomposition is mathematically represented as follows:

$$A_j[K]=\sum_{i=1}^{n}X[i]\times h[2K-i] \quad \text{and} \quad D_j[K]=\sum_{i=1}^{n}X[i]\times g[2K-i] \quad (3)$$

Where $A_j[K]$, $D_j[K]$ are respectively the outputs of low pass, h, and high pass, g, filters, at the level j ($j \in \{1,2,..L\}$). L is the total number of the levels of decomposition.

By using the concept of discrete wavelet transform, every profile is analyzed by applying three types of decomposition: the approximations, A, the details, D, and the profile of roughness, B, which is established by: $B_{i+1} = B_i + D_i$, where $B_1 = D_1$. The Figure 1 shows an example for this decomposition at levels 1 and 12. Various families of wavelet transform are available. In our implementation, three wavelet families are used (Symlets, Coiflets, and Meyer).

4. Variance analysis

By applying the three wavelet families (Symlets, Coiflets, and Meyer) with the three types of decomposition (A, B and D), each profile is decomposed in 9×L profiles where L is the total number of the levels of decomposition. Then the roughness average, R_a, is calculated. The main answer is then to determine if a R_a computed by a given wavelet W and a given decomposition R taken a scale ε discriminates the effect of a given process parameter and, more precisely, which scale of the wavelet transform will be the more appropriate. Consequently, a $R_a(W, R, \varepsilon)$ is obtained for all profiles. To process to this analysis, we decided to use the variance analysis. The main aspect is to analyze the experimental design by classical analyses of variance. This analyses is then performed at various scales ε and thus for the previously mentioned wavelets families and types of decomposition. The R_a equation is then defined as follows:

$$Ra(R,W,\varepsilon,K_1,K_2,...K_p,n)=\alpha(R,W,\varepsilon)+\sum_{j=1}^{p}\alpha_{K_j}(R,W,\varepsilon)+\sum_{j=1}^{p}\sum_{l=j+1}^{p}\beta_{K_j,K_l}(R,W,\varepsilon)+\xi_{K_1,K_2,...K_p,n}(R,W,\varepsilon) \quad (4)$$

Where $R_a(R,W,\varepsilon,K_1, K_2,..., K_p,n)$ is the value of the R_a parameter of the nth profile when process parameter of p are taken at the levels K_1, K_2,..., K_p, for an evaluation length ε. $\alpha K_j(R,W,\varepsilon)$ is the influence on the value of roughness parameter of the jth process parameter at the level K_j and $\beta K_j,K_l(R,W,\varepsilon)$ is the interaction between the j and l process parameters and $\xi K_1,K_2,...,K_p,n(R,W,\varepsilon)$ is a Gaussian noise with a null value.

For every evaluation length ε, wavelet family W and type of decomposition R, all coefficients of the models in Eq. 4 are computed thanks to the least square method.

To test the relevance of the proposed model, the variable F (Fisher-Snedecor) is the ratio of two chi-square distributions with degrees of freedom d_1 and d_2, respectively, where each chi-square has first been divided by its degrees of freedom, as a consequence, $F(\varepsilon,W, R, p)$ represents the effect of the p process parameter on the value of R_a computed on profile transform at the scale ε with a wavelet W and a decomposition R. For example, considering the pth process parameter using a wavelet W and a decomposition R, a greater F value at the scale ε_1 compared to the scale ε_2 (i.e. $F(\varepsilon_1,W,R,p) > F(\varepsilon_2,W,R,p)$) implies a greater relevance of the scale ε_1 to describe the effect of the pth process parameter.

However, this conventional statistical theory does not take into consideration the fact that a small variation in any score influences the value of the treatment index. That is why the variance analysis was combined with the Bootstrap theory. The aim of this statistical method is to generate N (N = 100) equivalent computational sets of data. Each set is generated from a corresponding experimental set of values by performing a permutation with replacement of these values.

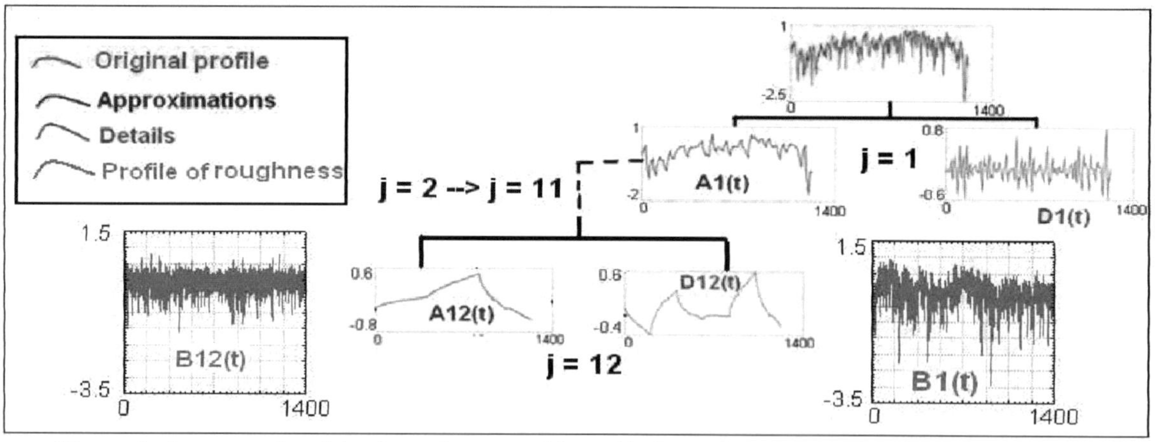

Figure 1: An example of the three decompositions (A, B, and D) for two levels, j = 1 and j = 12. The total number of the levels of decomposition is L = 17

5. Results

Firstly, we have to answer to the followings questions:

- Does the multiscale analysis allow us to find the spatial scale on which abrasion occurs?
- Does the scale of relevance depend on the various process parameters?

- Does the scale of relevance depend on the wavelet choice?
- Does the type of decomposition affect the relevance of the scale?

To answer to these questions, we decide to use the variance analysis previously described in the section 4 and computed the $F(\varepsilon_l, W, R, p)$ values for the different following set (Table 1):

Table 1. Experimental design (Tagushi) of belt finishing process and wavelets configuration (wavelets family, filtering method and resolution scale)

Factor	Factor Levels
Belt finishing process parameters	Belt feed, Hardness, Hardness versus Belt grid, Hardness versus Pressure, Belt grid, Belt grid versus pressure, Oscillation, Pressure, Rotation, and Time.
Wavelet Familty (W)	Symlets, Coiflets and Meyer
Filtering method	*A*, *B*, and *D*.
Resolution scale ε_l (μm)	20, 40, 80, 160, 320, 640, 1280, 2560, 5120, 10240, 20480, 40960, 81920, 163840, 327680, 655360, 1310720.

In a first time, we present the analysis of profiles obtained by the detail decomposition to visualize the relevance of the scale. Thus, we plot the value of F versus the scale ε for the three wavelets. Moreover the influence of each process parameter and three additional interactions were also studied. Figure 2 presents the various evolution of F for each choice of wavelet and process (or process interaction) parameter. It allows us to visualize the relevance of the scale and the effect of all process parameters. It is observed that curves are similar for the three wavelets for any process or process interaction parameter. To sum up, this clearly means that the choice of the wavelet does not influence the value of the most relevant scale. Moreover, this result seems to be verified independently of the choice of the process parameter or associated process parameter interaction.

As it can be observed on table 2, values of relevance do not depend on the wavelets.

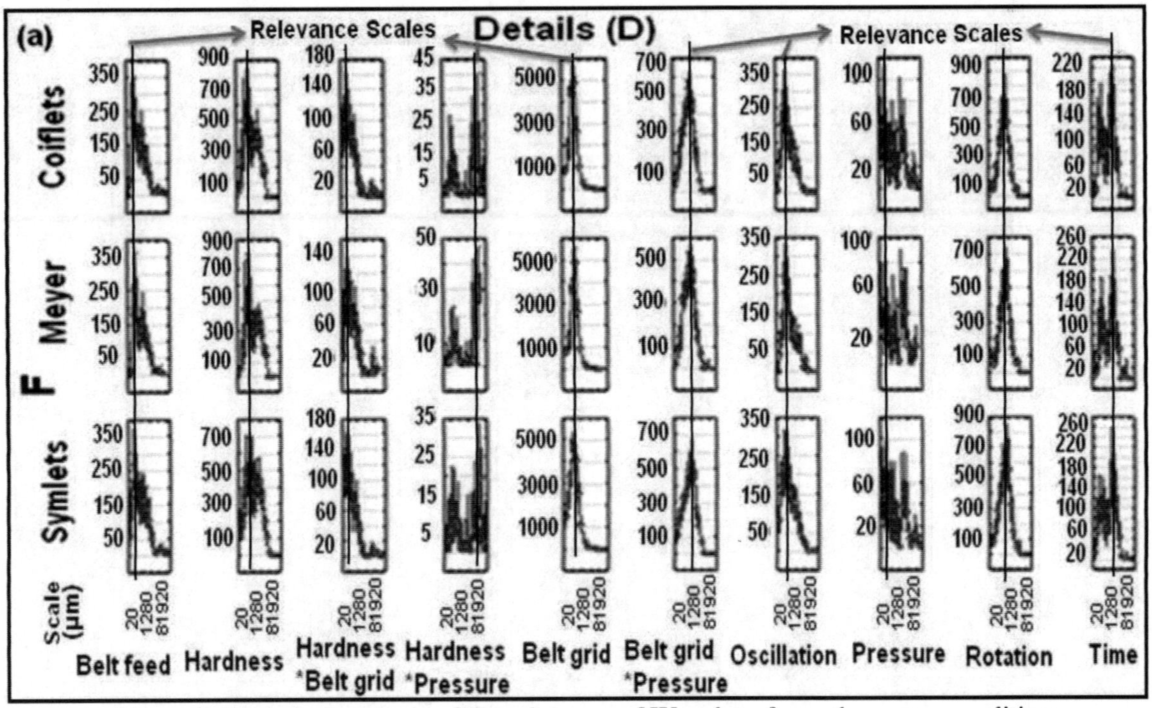

Figure 2: Statistical analysis of the relevance of Wavelets, for each process conditions

Table 2. Mean F value corresponding associated to the experimental design and their associated relevant scales (into bracket, in µm) with the three wavelet families.

	Belt feed	Hardness	Hardness* belt grid	Hardness* Pressure	Belt grid
Coiflets	**350**	**778**	**179**	**37**	**5927**
	[8,16]	[16,64]	[8,16]	[8192,32768]	[16,32]
Meyer	**364**	**878**	**135**	**50**	**6019**
	[8,16]	[16,64]	[8,16]	[8192,32768]	[16,32]
Symlets	**374**	**758**	**179**	**35**	**5609**
	[8,16]	[16,64]	[8,16]	[8192, 32768]	[16,32]
	Belt grid * Pressure	**Oscillation**	**Pressure**	**Rotation**	**Time**
Coiflets	**679**	**358**	**117**	**870**	**264**
	[32,128]	[8,16]	[8,64]	[64,128]	[256,512]
Meyer	**632**	**390**	**109**	**773**	**319**
	[32,128]	[8,16]	[8,64]	[64,128]	[256,512]
Symlets	**724**	**340**	**108**	**10860**	**284**
	[32,128]	[8,16]	[8,64]	[64,128]	[256,512]

6. Conclusion

Considering a Tagushi experimental design, the scale of relevance is the same for all the wavelets. Any wavelet leads to the same conclusion for this scale, independently of the roughness values. Moreover this allows us to find the most relevant scale range. These scales differ as a function of process or interaction criterion.

Our perspective is to test the relevance of the continuous wavelets and to apply this analysis to a larger database. The influence of roughness of each process or interaction process parameter on surface tribology property will then be studied.

References

[1] Khellouki A, Rech J and Zahouani H 2007 *Proc. Instit. Mech. Engin. Part B: J. Engin. Manuf.* **221** 1129.

[2] Rech J and A Moisan 2003 Le toilage: un moyen d'optimisation de l'intégrité des surfaces usinées, 16ème Congrès Français de Mécanique, Nice, France.

[3] Van Gorp A, Bigerelle M, El Mansori M, Ghidossi P and Iost A 2006 *Int. J. Mater. and Product Tech.*, **38(1)** 16.

[4] Vetterli M, Kovacevic J and Goyal V 2011 *The world of Fourier and wavelets: Theory, Algorithms and Applications*, (Attribution-NonCommercial-NoDerivs 3.0 Unported License., http://www.fourierandwavelets.org/book.pdf)

[5] Daubechies I 1992 *Ten Lectures on Wavelets* (2nd ed. Philadelphia: SIAM, CBMS-NSF regional conference series in applied mathematics 61).

[6] Torrésani B 1995 *Analyse continue par ondelettes* (InterEditions/CNRS Edition, Paris).

[7] Mallat S 1999 *A wavelet tour of signal processing* (2nd ed., Elsevier, Orlando).

13th International Conference on Metrology and Properties of Engineering Surfaces IOP Publishing
Journal of Physics: Conference Series **311** (2011) 012014 doi:10.1088/1742-6596/311/1/012014

Surface characterization based upon significant topographic features

J Blanc, D Grime, F Blateyron*
Digital Surf, 16 rue Lavoisier, F-25000 Besançon, France

E-mail: fblateyron@digitalsurf.fr

Abstract. Watershed segmentation and Wolf pruning, as defined in ISO 25178-2, allow the detection of significant features on surfaces and their characterization in terms of dimension, area, volume, curvature, shape or morphology. These new tools provide a robust way to specify functional surfaces.

1. Introduction

Surface texture has been focused on profile analysis for decades but recently areal instruments based on optical methods have become more and more popular as they allow non-contact measurements of rectangular surface portions. The new ISO 25178 contains several parts among which part 6 describes the accepted optical methods [1] that can be used to measure surfaces. Part 2 describes areal (3D) parameters [2] that are used to analyze surface texture. Apart classical parameters that are simple extensions of 2D equivalents (such as *Sa*, *Sq*…), the document defines new parameters that were designed to catch the complexity of modern surfaces.

Watershed segmentation used with a pruning criterion allows the creation of texture cells that can be characterized with parameters called *feature parameters*. These parameters make possible the specification of complex requirements on non-isotropic surfaces.

2. Watershed segmentation

While image segmentation algorithms have been used for a long time in image processing and vision systems, it is only recently [3, 4] that these algorithms have been introduced for surface measurements. A matrix representing surface heights is identical to a grey level

* Corresponding author.

image so the same tools can be applied almost without modification. However, one big drawback of segmentation is over-segmentation which detects features motifs everywhere in the image. Scott [5] introduced a discrimination method that makes it possible to merge small features into larger ones using a discrimination criterion developed by Wolf [6] and Kweon [7]. This so-called Wolf pruning can be controlled by a numerical value, for example 5% of Sz (peak-to-valley height) and therefore can be used as a specification operator in a drawing, in order to control surface function more accurately than with classical parameters.

3. Feature parameters

All 3D parameters in use before the new ISO 25178 standard, such as the set of 3D parameters defined in the SurfStand report [8] are "field parameters", i.e. they are calculated from all data points, by averaging, integration or Fourier transform. Almost all 2D parameters are also field parameters, except *RSm* and *Rc* [9] calculated from profile elements, and R&W parameters [10] calculated on roughness or waviness motifs.

A new family of parameters is introduced in ISO 25178-2, named "feature parameters" and defined as parameters calculated on selected topographic features, such as points, lines or areas, extracted by watershed segmentation of the surface.

Several topographic features are derived from definitions given 140 years ago by Maxwell [11] for the description of landscapes. Hills are surrounded by course lines created by the segmentation process (figure 1), and have extreme points (peaks). Symmetrically, dales are surrounded by ridge lines and have extreme points (pits). Course lines and ridge lines intersect at a saddle point (figure 2).

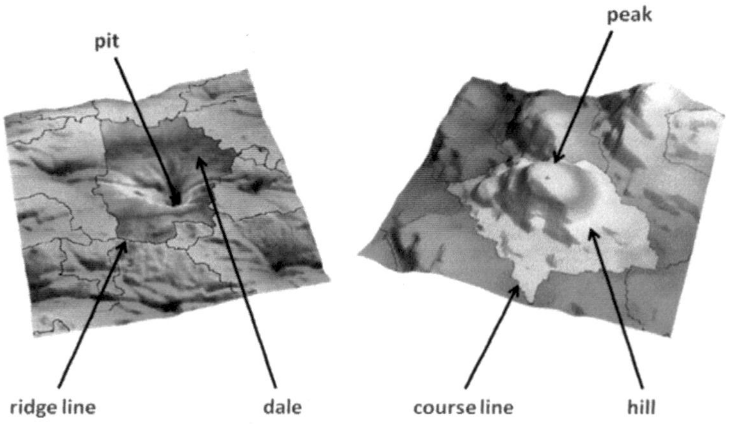

Figure 1. Topographic features on which Feature Parameters are calculated.

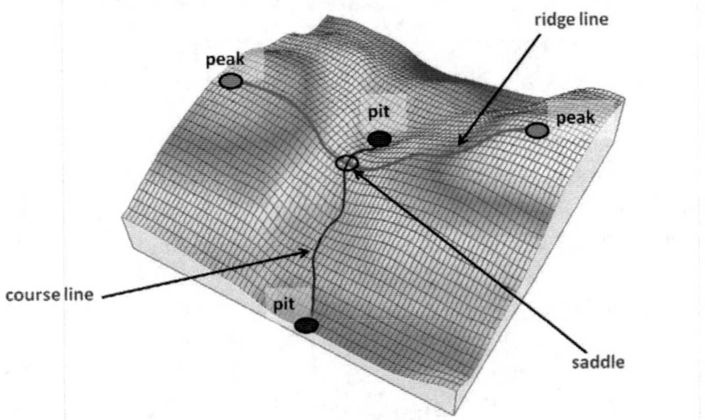

Figure 2. Saddle points are at the intersection of course and ridge lines

Hills and dales can be called "motifs" to keep using the vocabulary introduced with the R&W motifs that were calculated with a specific segmentation on profiles. Two new parameters provide the horizontal area (or projected area) of surface motifs, *Sda* on dales and *Sha* on hills. The volume of motifs is provided by two other parameters, *Sdv* and *Shv*.

Several other parameters can also be derived from surface motifs. The *developed area* of motifs, given in mm^2 or as a percentage when divided by the projected area can be interesting when used in conjunction with the motif height or the curvature of summits. The motif height is defined as the vertical distance between the extreme point and the closest saddle point (figure 3).

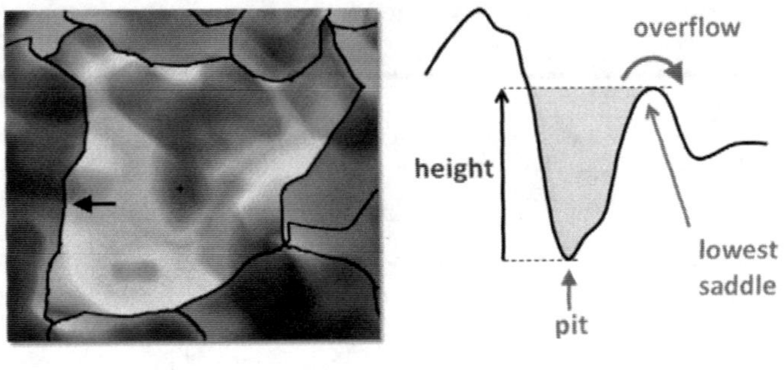

Figure 3. Definition of the motif height, illustrated on a dale.

For surfaces used in contact with other surfaces, segmentation provides a robust way to identify contact points with the significant peaks properly identified with the correct Wolf pruning. These peaks and surrounding hills can be characterized to predict wear and deformation, by calculating the surface curvature. The peak density (*Spd*) and the mean curvature of peaks (*Spc*) are useful in this context and these parameters greatly improve performances of previous *Sds* and *Ssc* parameters defined in the Surfstand report.

One of the most frequently used parameters on profiles is the *Rz* parameter. There were several versions of this parameter and other parameters (the ten-point height *Rz* still in use in the ASME B46.1 standard [12], the *R3z*, the *Rtm*, the *Ry*, or the last version of *Rz* defined in ISO 4287) were introduced with the same goal, to calculate a robust height. The segmentation method provides a robust way to identify significant peaks and valleys and therefore to calculate a robust surface height which is not sensitive to outliers with the *S10z* parameter as well as *S5p* and *S5v* (respectively peak height and valley height from the mean plane).

If we see surface motifs as texture cells, it is interesting to learn more about their morphology in terms of shape and dimension in the horizontal plane (i.e. without using the third dimension z). Several parameters commonly used in image processing [13] such as *form factor*, *aspect ratio* or *roundness* of motifs are calculated by ratios involving different types of diameters (*min, max, mean diameters, equivalent diameter*). They can be useful for discriminating between round and oblong motifs. The orientation of motifs, given by the direction of the largest dimension, can be used to separate grains or pores on heterogeneous surfaces.

4. Conclusion

Field parameters have been used for decades and provide good results on homogeneous surfaces. Today, more and more surfaces generated by modern fabrication methods are not isotropic or have structured cells to implement specific functional behavior. Feature parameters, calculated after a robust segmentation, are designed for the characterization of such modern surfaces. They can be seen as elements of a toolbox for engineers and metrologists that make it possible to define new specifications methods and obtain better correlation between parameter values and the functional requirements. Even though the watershed segmentation and the associated parameters were described many years ago, their use in the industry is still quite new.

Practical results and feedback for different applications are still to be published. In parallel, further work needs to be done on the definition of operands calculated on these motifs. Only a few parameters are proposed in the ISO standard, others are available in some analysis software packages [14] but a lot more can be done to define useful parameters.

5. 5. Examples

5.1. Automatic partitioning

Once detected and discriminated correctly, significant motifs can be used to partition surface structure either automatically using numerical criteria (e.g. motif area > value) or interactively by clicking inside a motif to select it (figure 4). For this application, the segmentation process is applied after a pre-filtering process such as a gradient or top-hat filter, in order to apply the segmentation on edges instead of hills and dales.

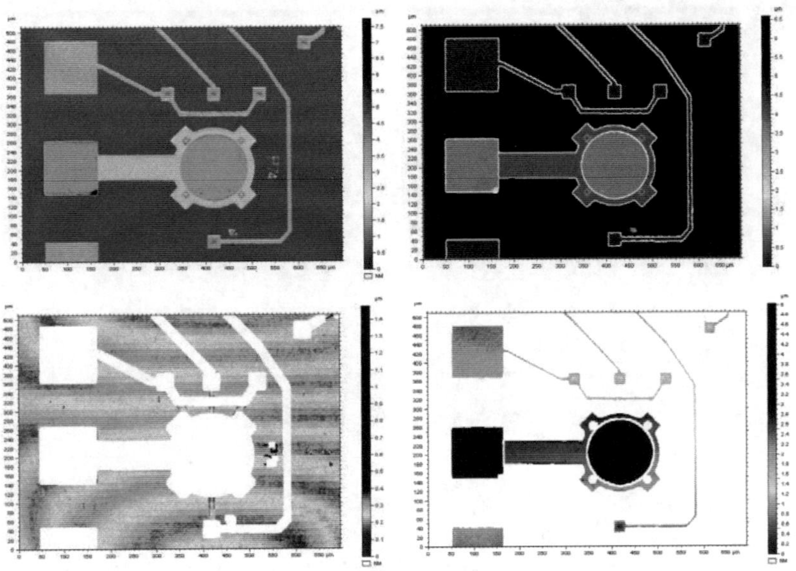

Figure 4. Automatic partitioning of a MEMS. Top left: original surface; top right: segmented features; bottom left: partitioned background; bottom right: segmented structures.

5.2. Texture morphology

Texture cells are calculated by segmentation and their dimensions are assessed. The mean value and the standard deviation of parameters are calculated over all motifs to provide a global evaluation. Parameters calculated on individual motifs can be tested against a tolerance and out-of-tolerance motifs can be identified and marked, or can be used to identify specific shapes in the surface.

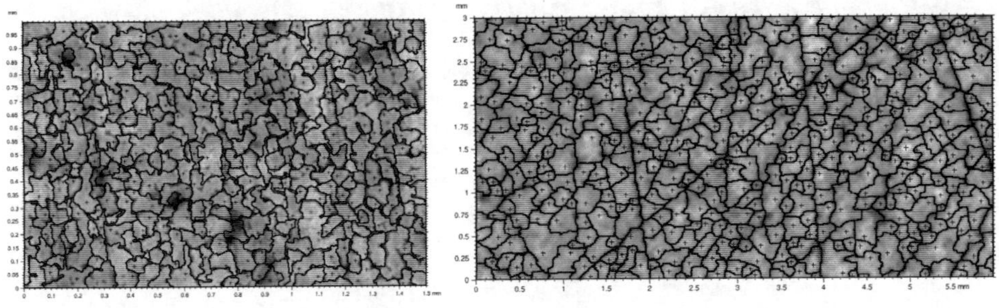

Figure 5. Left: Texture cells (motifs) detected on polished metal. Right: motifs calculated on a ceramic joint.

ISO 25178			
Feature Parameters			
Spd	16.3	1/mm²	*Density of peaks*
Spc	166	1/mm	*Arithmetic mean peak curvature*
S10z	157	µm	*Ten point height*
S5p	118	µm	*Five point peak height*
S5v	39.0	µm	*Five point pit height*
Sda	0.204	mm²	*Mean dale area*
Sha	0.0613	mm²	*Mean hill area*
Sdv	0.000418	mm³	*Mean dale volume*
Shv	0.000164	mm³	*Mean hill volume*

Figure 6. Feature parameters as per ISO 25178-2.

References

[1] ISO 25178-6 2010 *Classification of methods for measuring surface texture*
[2] ISO/FDIS 25178-2 2010 *Areal - Terms, definitions and surface texture parameters*
[3] Scott P J 1992 The Mathematics of motif combination and their use for functional simulation *Int J. Mech. Tools Manufact.* **32** 69-73
[4] Barré F, Lopez J 2000 Watersheds lines and catchment basins: a new 3D-Motifs method *Int. J. Mech. Tools Manufact.* **40** 1171-1184
[5] Scott P J 2004 Pattern Analysis and Metrology: The extraction of stable features from observable measurements *Proc. R. Soc. Lond. A* **460** 2845-2864
[6] Wolf G 1991 A Fortran subroutine for cartographic generalization *Computers & Geoscience* **17.10** 1359-1381
[7] Kweon I S, Kanade T 1994 Extracting topographic terrain features from elevation maps *CVGIP: image understanding* **59.2** 171-182
[8] Blunt L, Xiangqian J 2003 *Advances techniques for assessment surface topography* Kogan Page Science ISBN 1-9039-9611-2
[9] ISO 4287 1997 *Profile method - Terms, definitions and surface texture parameters*
[10] ISO 12085 1996 *Profile method – Motifs parameters*
[11] Maxwell J C 1870 On hills and dales *Philosophical magazine* **II** 233-240
[12] ASME B46.1 2002 *Surface texture (surface roughness, waviness and lay)*
[13] Russ J C 1995 *The image processing handbook* CRC Press ISBN 0-8493-2516-1
[14] MountainsMap™ surface analysis package, www.digitalsurf.com

3D-characterization method and morphological filtering for the assessment and the design of friction optimized surfaces

S Gröger[1,1], T Burkhardt[2] and M Dietzsch[1]
[1]Chemnitz University of Technology, Institute of Production Measuring Technology and Quality Assurance, Reichenhainer Str. 70, 09126 Chemnitz, Germany
[2]Fraunhofer Institute for Machine Tools and Forming Technology IWU, Chemnitz, Germany

E-mail: sophie.groeger@mb.tu-chemnitz.de

Abstract. For a specific manipulation of friction surfaces it is important to measure and calculate geometrical parameters to derive the tribological behavior. The new functional approach presented in this paper is the calculation of the characteristic lateral extension of the real contact surface as well as the representative contact radius by applying morphological filters to a 3D-set of data. All surface characteristics, including form, waviness, roughness as well as defined microstructures, are extracted holistically with a 3D Coordinate Measuring Instrument or a Form Measuring Instrument, but with the smallest available tip radius. The paper presents the benefit of this holistic extraction method and the application of morphological filtering for the description of the contact form (plateau or sphere), the real contact surface, number of contacts, the typical contact radius and the typical lateral extension of the micro contact plateaus.

1. Introduction

The control of friction is very important for an energy-efficient function of mechanical motion. The focus of current research is set on the optimization of modern lubricants as well as on the design of friction optimized surface structures. One approach is the development of geometric determined microstructures with equally spaced microcavities. The other way is the improvement of surfaces by defining the contacting area. The intention of the following explanation is therefore not only to define friction optimized surfaces with microstructures but also to assess common friction surfaces without microstructures.

Microstructures are preferably manufactured with erosive or cutting technologies. The application specific design and dimensioning of such microstructures are still empirical. For a better understanding of the mechanisms of a certain surface quality or a defined, determined surface structure for tribological behavior, a detailed interpretation of the effects and mechanisms in and on microstructures is needed. Two types of microstructures are discussed: Microstructured surfaces Type 1 and microstructured surfaces Type 2 [1].

[1] Corresponding author. Tel.: +49.371.531.32212; fax: +49.371.531.832212.
E-mail address: sophie.groeger@mb.tu-chemnitz.de (Sophie Gröger)

Microstructured surfaces Type 1 are characterized by cavities with defined geometrics equally spaced on the surface. The friction mechanisms primarily occur on the remaining plateau. Within these microcavities, lubricants or particles, which could cause more friction or abrasion, can be stored (figure 1). The bottom part of this figure zooms in on the dependency of the lubrication film thickness and the geometrical characteristics compared to a zoomed in explanation for a surface without microstructures (Figure 1 bottom) over the friction length. Without microstructures the lubrication film thickness decreases over the friction length. With microstructures (figure 1) a periodical regeneration of the lubrication film can be achieved by phases of contact (contact length) and non contact (length without contact). For these secondary mechanisms, the lateral dimension and the volume of the microcavities as well as the surface related part of the nominal surface become important. Besides the lateral dimensions of the microcavities, the dimensioning of its depth structure, depending on the sliding speed and lubricant viscosity, is also important to create hydrostatic pressure.

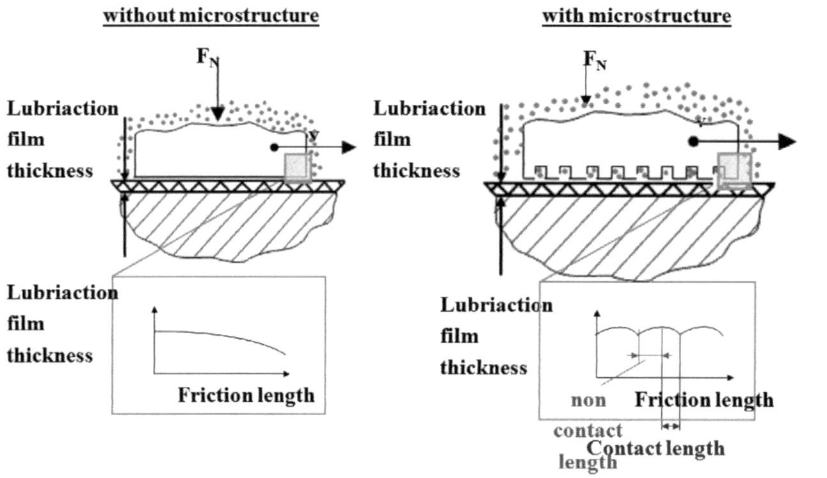

Figure 1. Mechanisms of surfaces without microstructure; and mechanisms of microstructured surfaces Type 1.

Microstructured surfaces Type 2 are three dimensional geometries without cavities. A simple example is the surface presented in figure 2.

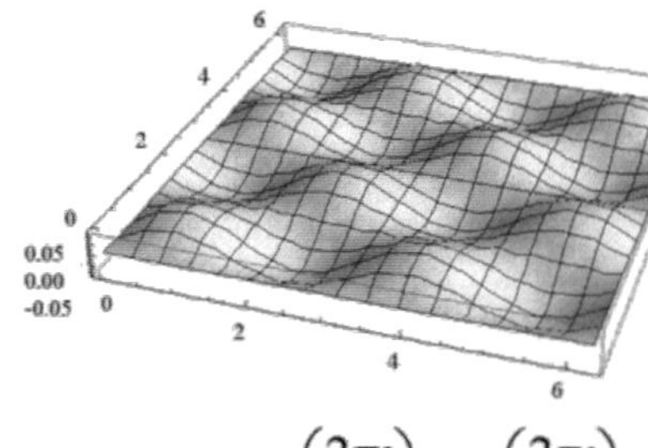

$$z = a \cdot \sin\left(\frac{2\pi x}{L}\right) \cdot \sin\left(\frac{2\pi y}{L}\right)$$

z(x,y) height at position (x,y)
A amplitude
L period length

Figure 2. Geometrical characteristics of Type 2 [1].

It can be shown that material stress p_0, caused by an applied nominal contact pressure p_N only depends on the quadratic ratio of the structure depth Z = 2a and the period length L.

$$p_0 = \sqrt[3]{3\pi \frac{Z^2}{L^2} \frac{p_N E^2}{(1-\upsilon^2)}} \tag{6}$$

Looking at the single contacts as „Hertzian Contacts,“ the lateral dimension of adhesion related contact areas, the material stress and hydrodynamic friction behaviour can be optimized application specifically with the parameters Z and L. These simplified mechanisms are long known but not implemented in production processes. This might be caused by missing measuring methods as well as manufacturing processes. Today there is no general strategy to derivate the parameters under practice conditions.

2. Assessment and design of friction optimized surfaces

Today the geometrical properties are separated into form, waviness and roughness [2]. Different measuring instruments and measuring methods are used to extract these properties independently. A correlation of the geometrical properties and the functional characteristics is not possible with the separated information of the surface. The real contact surface of a workpiece interacts with the counterpart and is influenced by the combined effects of the contact form, number of contacts, the typical contact radius and the typical lateral extension. The understanding of this interaction needs a holistic view of the surface and its properties. Holistic means that all geometrical properties are assessed from one measurement to maintain their connection.

2.1. 3D characterization method

There are different ways to extract a workpiece in a holistic way. The easiest way is to use computer tomography. The problems with that are still the limited resolution and the definition of the electro-magnetic surface. Therefore other methods have been developed at the Chemnitz University of Technology. They especially consider the mechanical surface as the functional surface for tribological functions. In ISO standards the mechanical surface is defined for the tactile measuring method and is one functional approximation of the real surface.

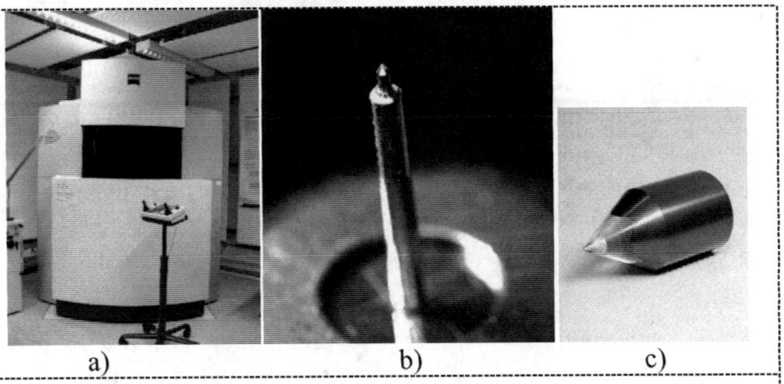

Figure 3. (a) 3D-CMM F25; (b) CMM-probe with tip radius 2 µm; (c) stylus tip mounted on the probing shaft.

Mechanical surfaces can be extracted holistically with a 3D Coordinate Measuring Instrument (3D-CMM) or a Form Measuring Instrument, but with the smallest available tip radius. The tip radii used are 2 µm and 5 µm. Common 3D-CMMs do not have tip radii like that. Figure 3 (b) therefore shows a new developed probe for the 3D-CMM F25 by Zeiss (figure 3 (a)). The stylus tip with a radius of

2 μm (figure 3 (c)) is mounted on the special probing system for the F25. The low tracing force of this probing system allows this new development. This approach is still in the testing period. Measuring results will be published in the near future.

A second approach is the holistic extraction with a Form Measuring Instrument MFU 100 by Mahr. Flat and cylindrical surfaces can be measured holistically in one setting with a tip radius of 5μm. For example circular profiles are extracted by rotating the table and shifting the stylus after one measurement to the next position (figure 4 (a)) over the complete flat ring surface of about 12 mm. Figure 4 (b) shows the extraction result. The real contacting surface is limited by the form deviation if the counterpart is perfectly flat. The zoomed detail of the surface plot (figure 4 (c)) reveals the applied microstructure. The extracted profile above figure 4 (c) (position marked with the dotted line) reveals the sinus type surface. Sampling distance and length depend on the characteristic and the dimension of the microstructure and the workpiece.

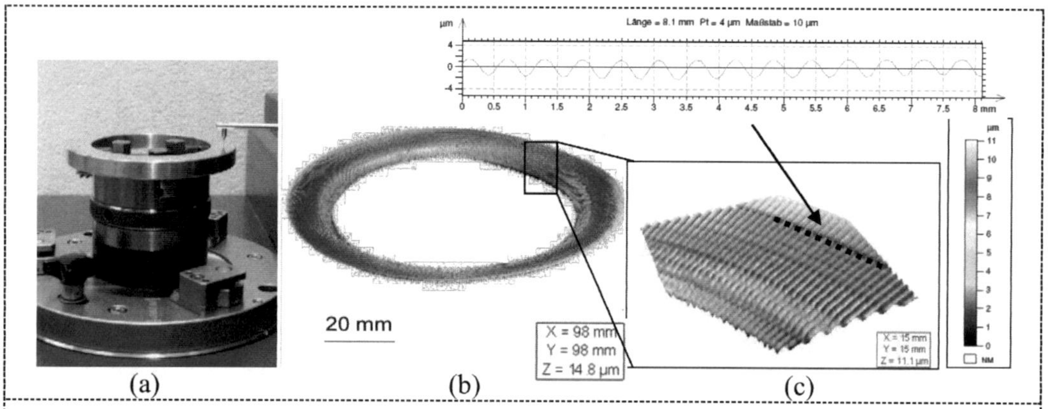

Figure 4. (a) Measurement with a Form Measuring Instrument (b) 3D surface plot; (c) zoomed detail of the extracted surface.

There needs to be further research to develop these approaches into a general method. The bases for this method and the connection of the functional geometry of the workpiece to the probing conditions are surface parameters to describe the functional behaviour.

2.2. Design of friction optimized surfaces and the development of new parameters

In the past couple of years a set of 3D parameters for roughness has been developed. It is defined in ISO/DIS 25178 part 2 [3]. The parameters are separated in height, spacial, hybrid, function and related parameters as well as feature characterization. These parameters are not sufficient for the design of friction optimized surfaces. The mean line filtering to separate the form and the statistical character limits the holistic description of the surface.

For a couple of years the descriptions of surfaces, with the application of morphological filters, have been qualified at the Chemnitz University of Technology [4]. Figure 5 describes the basic morphological filtering operations. The calculated mean line of a structuring element represents a dilatation. This is the same as extracting the surface with a stylus with a spherical tip and calculating the line of the centre points. The reversal of this is the erosion. Erosion applied after dilatation leads to the reconstructed mechanical surface and is also defined as closing. Algorithms are available to simulate different structuring elements on a surface to derive different closing surfaces. The other possibility is to simulate erosion before dilatation as an opening operation. This is possible for 2D and 3D surface data [5]. Closing and opening surfaces describe the resulting surface after applying the morphological filter operation.

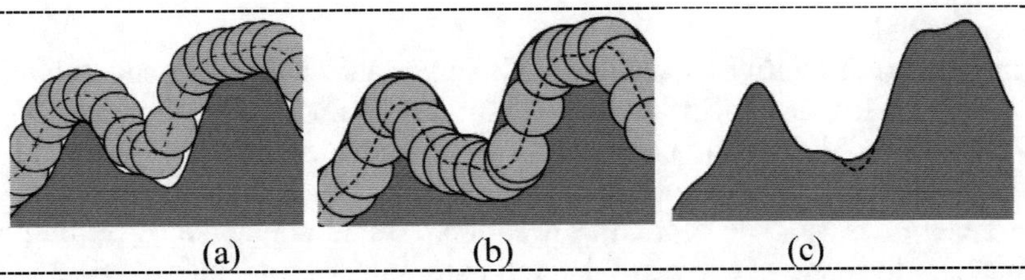

Figure 5. Morphological Operations (a) dilatation; (b) erosion; (c) reconstructed mechanical surface [5].

With these closing operations the extracted surface can be separated in form, waviness and roughness portions but in connection to each other by a common datum system. This datum represents the counterpart and is, in the first approach, an ideal flat or cylindrical feature. Furthermore distances and amplitude parameters as well as void and material volume can be calculated as 3D parameters to associate geometrical surface characteristics to tribological parameters (figure 6). Supporting points are a set of points that have the same x- and y- coordinate for different closing surfaces. These new parameters based on the common datum system will be implemented in evaluation software soon.

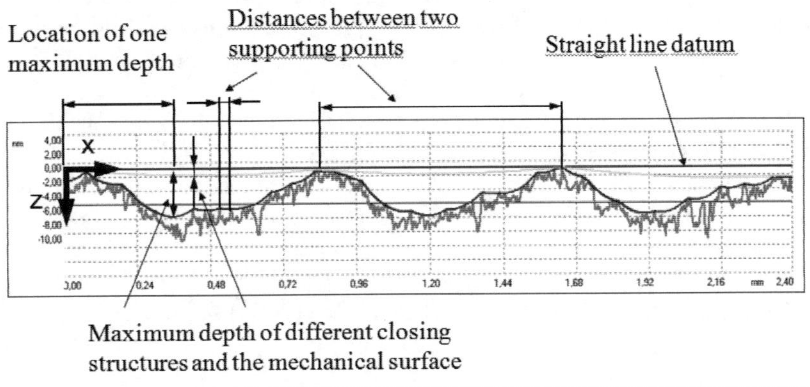

Figure 6. Common datum system

With morphological opening operations the real radius of a surface can be calculated (figure 7). A more or less good approximation of the simulated surface to the mechanical surface can be achieved by simulating closing surfaces with different radii for the structuring element. No difference between the mechanical surface and the simulated surface is an indicator for the best possible approximation.

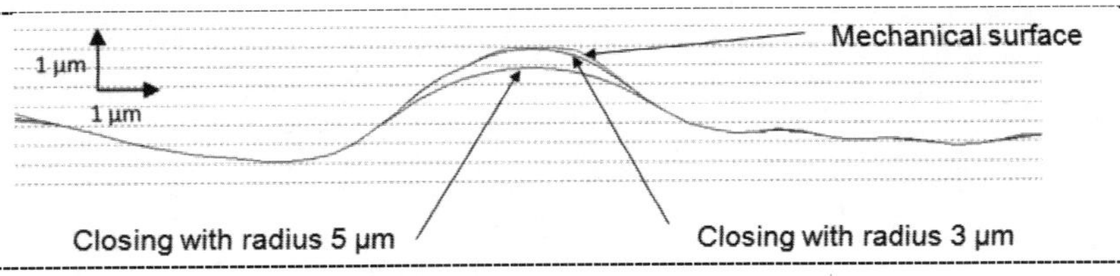

Figure 7. Radius simulation

Algorithms have to be developed to calculate the distribution of contacting radiuses as one influencing value to characterize the tribological behaviour. The distribution of volume parameters, number of contacts and the lateral extension of the micro contact plateaus have to be visualized.

3. Summary and outlook

For a specific manipulation of friction surfaces it is important to measure and calculate geometrical parameters to derive the tribological behavior. All surface characteristics, including form, waviness, roughness as well as defined microstructures have to be extracted holistically with a 3D Coordinate Measuring Instrument or a Form Measuring Instrument, but with the smallest available tip radius. A common datum system has been developed to describe the real contact surface, number of contacts, the typical contact radius and the typical lateral extension of the micro contact plateaus. With morphological operation a segmentation of the surface based on this common datum system is possible.

The future intention of the presented research is the development of a design guide for friction optimized surfaces with microstructures and an optimization tool for common friction surfaces to support designers and application engineers. For an improved description of the tribological behaviour the new characterization method and the newly developed surface parameters will be applied to different workpieces not only with microstructured surfaces. The designer will be assisted with tolerance principles, 3D-notation rules, parameter tables, and so on.

4. References

[1] S. Gröger, T. Burkhardt, U. Eckert, C. Kühn: New Characterization Methods for the Assessment and the Design of Friction Optimized Surfaces, International Chemnitz Manufacturing Colloquium, Verlag Wissenschaftliche Scripten, 2010, p. **839-846**

[2] DIN 4760: 1982. Gestaltabweichungen, Begriffe

[3] ISO/DIS 25178-2:2007: Geometrical product specifications (GPS) - Surface texture: Areal - Part 2: Terms, definitions and surface texture parameters

[4] S. Gröger; M. Dietzsch; M. Gerlach: A New Envelope System for the Evaluation of Functional Surfaces. XII. International Colloquium on Surfaces: Proceedings, 1/2008, Aachen: Shaker Verlag, 2008, **50-59**

[5] M. Krystek: Morphological filters in surface texture analysis. Proceedings of the XI. International Colloquium on Surfaces ICS, Chemnitz 1(2004) **43-55**

Journal of Physics: Conference Series **311** (2011) 012016 doi:10.1088/1742-6596/311/1/012016

A study on the frictional response of reptilian shed skin

H A Abdel-Aal[1]*, R. Vargiolu[2], H. Zahouani[2], M. El Mansori[3]

[1] Arts et Métier ParisTech, Rue Saint Dominique BP 508, 51006 Chalons-en-Champagne, France
[2] Laboratoire de Tribology et Dynamique des Systèmes, UMR CNRS 5513, ENI Saint Etienne - Ecole Centrale de Lyon –36 Avenue Guy de Collongue, 69131 Ecully cedex. France
[3] Ecole Nationale Supérieure d'Arts et Métiers, 2, cours des Arts et Métiers - 13617 Aix en Provence cedex 1, France

Hisham.abdel-aal@ensam.eu

Abstract. Deterministic surfaces are constructs of which profile, topography and textures are integral to the function of the system they enclose. They are designed to yield a predetermined tribological response. Developing such entities relies on controlling the structure of the rubbing interface so that, not only the surface is of optimized topography, but also is able to self-adjust its tribological behaviour according to the evolution of sliding conditions. In seeking inspirations for such designs, many engineers are turning toward the biological world to study the construction and behaviour of bio-analogues, and to probe the role surface topography assumes in conditioning of frictional response. That is how a bio-analogue can self-adjust its tribological response to adapt to habitat constraints. From a tribological point of view, Squamate Reptiles, offer diverse examples where surface texturing, submicron and nano-scale features, achieves frictional regulation. In this paper, we study the frictional response of shed skin obtained from a snake (Python regius). The study employed a specially designed tribo-acoustic probe capable of measuring the coefficient of friction and detecting the acoustical behavior of the skin in vivo. The results confirm the anisotropy of the frictional response of snakes. The coefficient of friction depends on the direction of sliding: the value in forward motion is lower than that in the backward direction. Diagonal and side winding motion induces a different value of the friction coefficient. We discuss the origin of such a phenomenon in relation to surface texturing and study the energy constraints, implied by anisotropic friction, on the motion of the reptile.

1. Introduction

Reduction of friction in moving assemblies is key to energy savings. This is because it reflects on the quantity of energy spent to combat frictional resistance to motion. Reduction of friction depends on many factors. One of the important factors is the topography of the rubbing interfaces. Custom engineering of the texture and topography of a rubbing interface, therefore, have acquired momentum within the last decade or so. To this effect many researchers envision a so called "deterministic surfaces" as the future of surface engineering. These surfaces not only would be of optimized topography but also their function would be an integral feature of the overall function of the system they enclose.

Construction of structured deterministic surfaces depends on understanding the interaction between surface topography, roughness parameters and frictional response. Structuring of surfaces in tribology has multiple goals. For lubricated surfaces it may be desirable to alter the topography in a manner that establishes a full hydrodynamic regime in within a short distance of sliding so that complete separation between surfaces is achieved. Controlled adhesion may also be a goal of surface structuring. Additionally, constructing a self cleaning surface or a hydrophilic or hydrophobic surface may be one of the required goals. In all, a structured surface for enhanced tribological performance should ideally be capable of tuning its frictional response to optimize system function. Achieving these goals is not necessarily easy, to date it may be fairly stated that generation of structured surface patterns is an art. In nature there is an abundance of examples of hierarchically structured naturally occurring surfaces that deliver super functionality. The diversity and richness of these examples have fuelled interest in the study of natural surface designs with the goal of mimicking them or deducing surface design rules for deterministic constructs. Of particular interest is investigation of the interaction between surface topography and friction response; and the topological design features of the surface that allow a natural surface to optimize its' frictional behavior in response to changes in the surroundings or in the contacting terrain. A rich resource for such a purpose is frictional behaviour of squamata.

Squamata, or the scaled reptiles, comprises the largest order of reptiles. It includes lizards and snakes. Squamate reptiles number approximately 8000 living species and are a major component of the world's terrestrial vertebrate diversity. They are the most variably-sized order of reptiles, ranging from the 16 mm Jaragua Sphaero (Sphaerodactylus ariasae) to the 8 m Green Anaconda (Eunectes murinus). Their evolutionary history dates back to the early Jurassic or late Triassic, they have diversified on all major continents, and they occupy a remarkable diversity of ecological niches [1–5]. For sliding, the study of legless-locomotion-reptiles, i.e., snakes, is technically attractive.

Snakes have multi-modes of motion (slithering, crawling, serpentine movement etc). Transfer of motion between the body of the snake and the substrate depends on generation of sufficient tractions. Generation of traction and accommodation of motion is handled through the skin. Thus the skin of the snake transfers frictional traction transfer and accommodates the energy consumed in facilitating motion. Motion is initiated through muscular activity that comprises sequences of contraction and relaxation of appropriate muscle groups. The number, type and sequence of muscular groups responsible for the initiation of motion, and thus-employed in propulsion, will vary according to the particular mode of motion. It will also depend on the habitat and the surrounding environment. This also will affect the effort invested in initiation of motion and thereby also affects the function of the different parts of the skin. So that, in general, different parts of the skin will have different functional requirements. Moreover, the life habits of the particular species (e.g., defense, hunting, and swallowing) will require different topographical features within the different parts of the skin. These aspects render snakes ideal study objects when the role of functional surface in controlling friction is considered.

In this work, therefore, we report on the dynamical frictional behavior of shed skin of a snake (Python regius). The aim of the study is to characterize the frictional response of the ventral scales both in linear and diagonal motion.

2. Background

2.1. Structure of Snake skin

Snakes are entirely covered with scales of various shapes and sizes. Scales protect the body, aid in locomotion, help retain moisture, and alter the surface characteristics. Snake scales are formed by the differentiation of the snake's underlying skin or epidermis. Each scale has an outer surface and an inner surface. A snake hatches with a fixed number of scales. The scales do not increase in number as the snake matures nor do they reduce in number over time. Snakes periodically molt their scaly skins and acquire new ones. The scales however grow larger and may change shape with each molt.

Reptilian skin has two principal layers: the dermis which is the deeper layer of connective tissue with a rich supply of blood vessels and nerves; and the epidermis, which in reptiles consists of up to seven sub-layers or "strata" of closely packed cells, forming the outer protective coating of the body. The "epidermis" has no blood supply, but its' inner most living cells obtain their nourishment by the diffusion of substances to and from the capillaries at the surface of the "dermis" directly beneath them. There are seven epidermal layers, the "stratum germinativum", the deepest layer lining cells which have the capacity for rapid cell division and the six layers which form each "epidermal generation", the old and the new skin layers. These are: the clear layer and the lacunar layer, which matures in the old skin layer as the new skin is growing beneath; the (α)–layer, the mesos layer and the (β)-layer, these layers consist of cells which are becoming keratinized with the production of two types of keratin (α and β keratin). These cells are thus being transformed into a hard protective layer. Finally, there is the "oberhautchen" layer, which forms the toughest outer most layer of keratinized dead skin cells. The shedding of the old epidermal layer of the skin generally occurs as a single piece (molt) after a complex process. The shed epidermis of snake skin consists of four layers: the outermost Oberhautchen and the β-layer (mainly protein), the mesos layer (lipid-rich) and the inner α-layer.

2.2. *Characterization of skin.*

All observations reported herein pertain to shed skin obtained from a 115 cm, 14 years old male Ball Python (Python Regius) housed individually in a glass container with news paper substrate. We identified 12 (twelve) positions on the reptile hide for observation. Figure 1 depicts the general features of the skin. Dorsal skin is composed of several dark and light coloured blotches, whereas the ventral skin is mainly cream in colour with occasional black markings. Note that the micro features of the scales comprise fibrils of the size of a few microns. These are of in-consistent shape and spacing. Fibril tips point toward the tail, within the dark-coloured dorsal scales; fibrils are tapered and have a sharp tip. Scales within the bright coloured and the ventral regions have a more rounded tip and appear to be of uniform width throughout the fibril length. Moreover the density of the fibrils seems to be different within the different colour regions (denser within the dark coloured region). Since ventral fibrils are thought to be the origins of the frictional response of the reptile, we undertook further analysis of their geometry. The results are shown in figure 2 (a and b).

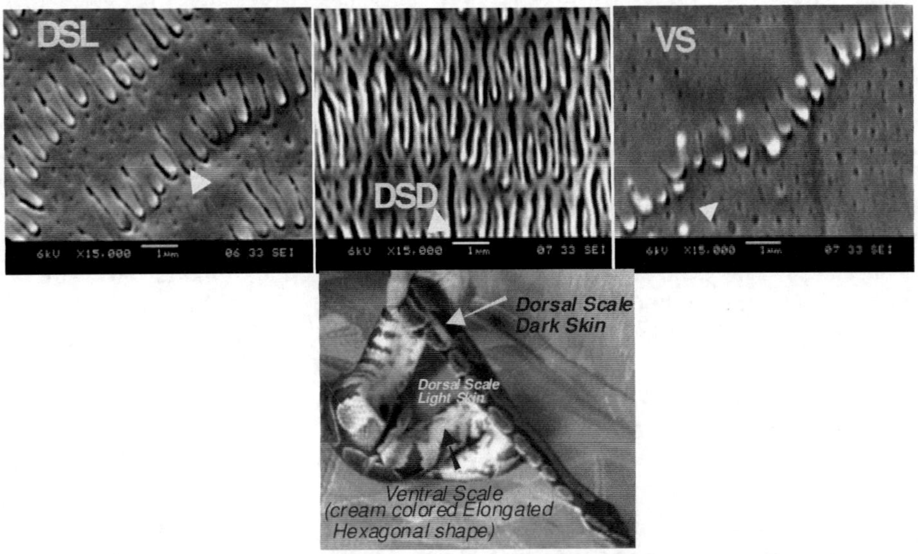

Figure 1. General appearance of the Python regius and SEM details of the skin at three positions: Dorsal skin light colored (DSL), Dorsal skin dark colored (DSD) and the ventral skin (VS). Magnification x=15,000. Scale marker 1μm.

3. Friction behaviour

All measurements utilized a patented tribometer that includes a tribo-acoustic probe sensitive to the range of friction forces and the acoustic emission generated during skin friction. The device is capable of measuring normal, tangential loads and also of detecting sound emission due to sliding. Detailed description of the tribometer is given elsewhere [6]. The skin used in measurements consisted of 150 mm long patches taken from four locations on the shed hide. Skin samples didn't receive any chemical or physical treatment. For each patch, measurements were taken for the dorsal and ventral sides of the skin. Four directions of sliding were used: straight forward (SF), straight backward (SB), diagonal forward (DF), and Diagonal backward (DB). Figure (3-a) depicts the directions used to obtain the measurements and their orientation with respect to the head and tail of the species.

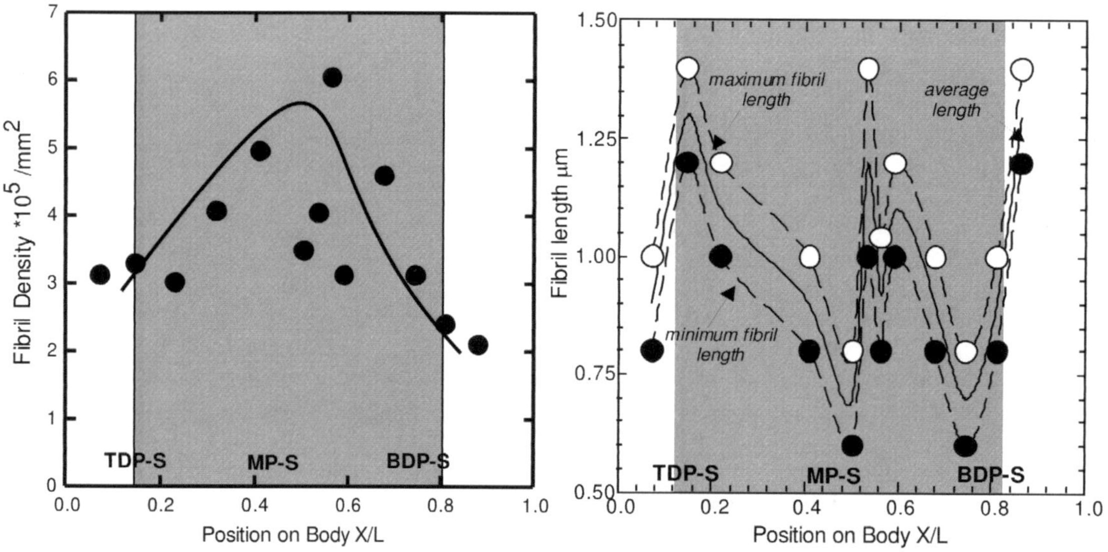

Fig. 2. Distribution of the fibril density, left hand side, and fibril length, right hand side, along the body of the reptile

Figure (3-b) presents a summary plot of the measured COF. Each value is an average of five consecutive measurements taken at the same direction of sliding and identical loading conditions. The results imply that resistance to forward motion is, in general, less than that for moving backwards, observe the difference between the value obtained for the SF direction and that for the SB direction. Anisotropy of the COF is also noted from the results. The COf when moving diagonally is different from that in straight motion. Resistance to motion, however, from tail to head (i.e., forward) is less than that from head to tail (i.e., backward) regardless of the direction of motion (forward or diagonal). These general trends are in line with measurements obtained by other researchers [7, 8] who confirmed the anisotropy in friction as well as the increase in resistance to backward motion (although on different species). However, in this work, dynamic measurements of the COF in the diagonal directions are reported for the first time in literature.

The origin of the frictional behaviour is rooted in the structure of the surface of the snake. In particular, the presence of fibrils within the scales aids the beast in conditioning its frictional response. The fibrils are asymmetric. The slope of fibril tips is gradual in the direction from head to tail and steep from tail to head. This asymmetrical tip-shape offers directional resistance to motion through acting as a ratchet which offers less resistance to the motion of the beast in the forward linear direction than that offered in the reverse direction.

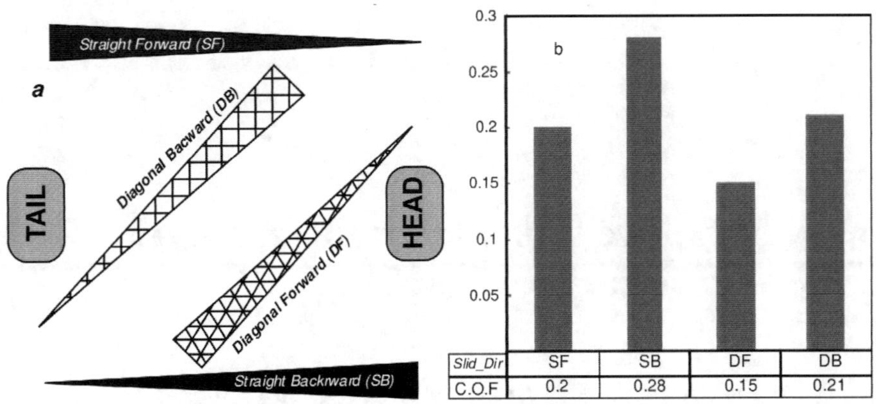

Fig. 3 (a) Schematic illustration of the directions of sliding for which coefficient of friction measurements were recorded (b) coefficient of friction values for each of the sliding directions illustrated in figure 3-a.

The overall structure of the ventral scales affects friction. The connector tissue which links the scales acts as a spring which flexes when the animal attempts to move backwards. Flexing dissipates additional locomotion energy generated by the beast. This renders backward motion infeasible as it becomes cost prohibitive in terms of energy expenditure.

It is likely that the trends exhibited by the COF (anisotropy) are also a function of the chemical composition of the Oberchautchen layer. This layer is mainly composed of β keratin, which is lamellar. Other natural materials such as camel and horsehair, which have the same keratin, composition are known to exhibit a so-called differential friction effect (DFE). In DFE the frictional work required by a fibre to slide over another fibre is greater in the direction of tip-to-root than the converse. However in the absence of additional data a conclusion cannot be reached.

Conclusions

We presented a study of the frictional characteristics of reptilian shed skin (Python regius). Results showed the directional dependency of the COF. In the forward motion (Tail-to-Head) the COF was less than that measured in the opposite direction (Head-to-tail). A similar trend was reflected in the values obtained for the COF in diagonal sliding.

References

[1] L.J. Vitt, E.R. Pianka, W.E. Cooper Jr. and K. Schwenk, History and the global ecology of squamate reptiles, Am. Nat. 162 (2003), pp. 44–60.

[2] Cheng Chang, Ping Wu, Ruth E. Baker, Philip K. Maini, Lorenzo Alibardi, Cheng-Ming Chuong, Reptile scale paradigm:Evo-Devo, pattern formation and regeneration, Int. J. Dev. Biol. 53 (2009) 813-826

[3] Alibardi, L., Thompson, M. B. Keratinization and ultrastructure of the epidermis of late embryonic stages in the alligator (Alligator mississippiensis). J. Anat. 201 (2002) 71-84.

[4] Ruibal R The ultrastructure of the surface of lizard scales.Copeia (1968) (4):698–703

[5] Chiasson RB, Bentley DL, Lowe CH Scale morphology in Agkistrodon and closely related crotaline genera. Herpetologica 45 (1989) 430–438

[6] Zahouani, H., Vargiolu, R., Boyer, G., Pailler-Mattei, C., Laquièze, L., Mavon, A., Friction noise of human skin in vivo Wear, 267 (5-8), (2009), 1274-1280.

[7] Hazel, J., Stone, M., Grace, MS, and Tsukruk, VV Tribological design of biomaterial surfaces for reptation motions. Polymer Preprints 39(1998) 1187-1188.

[8] J. Hazel , M. Stone, M. S. Grace and V. V. Tsukruk, Nanoscale design of snake skin for reptation locomotions via friction anisotropy, Journal of Biomechanics, 32, 5, (1999) 477-484.

Journal of Physics: Conference Series **311** (2011) 012017 doi:10.1088/1742-6596/311/1/012017

Efficiency of adaptive sampling in surface texture measurement for structured surfaces

J Wang[a], X Jiang[a], L A Blunt[a], R K Leach[b] and P J Scott[a]

[a] Centre for Precision Technologies, School of Computing and Engineering, University of Huddersfield, Queensgate, Huddersfield HD1 3DH, UK

[b] Engineering Measurement Division, National Physical Laboratory, Hampton Road, Teddington, Middlesex TW11 0LW, UK

Email: j.wang4@hud.ac.uk

Abstract. Adaptive sampling has been used as an efficient sampling strategy in metrology for many years. However, most of the research into adaptive sampling has concentrated on the area of form measurement by coordinate measuring machines. This paper will discuss the use of adaptive sampling, primarily to reduce the measurement time when making high density areal surface measurements, for instruments that measure surface texture using scanning mechanisms, such as stylus profilometers. Specifically, the paper will concentrate on a common adaptive sampling method known as indirect sampling and a modified version is proposed in this paper. Simulated sampling analyses and three typical micro-scale structured surfaces (a vee-groove-like surface, a rectangularly tessellated micro-lens array and a MEMS device surface) are used as test cases. Using the tensor product order 2 B-spline reconstruction, the root mean square deviation from the original surface is calculated. By comparison with the widely used uniform sampling technique, the performance analysis results show that the modified adaptive sampling has advantages by improving the measurement accuracy, reducing the data size and reducing the measurement duration.

1. Introduction

In surface measurement, a part of the measuring error comes from the sampling process in which the continuous surface is digitised. Although Shannon's sampling theory [1] states that a band-limited signal can be completely reconstructed from its discrete samples, exact band-limited signals do not usually exist in practice. For this reason, information about the un-sampled regions is inevitably lost. Some intelligent sampling designs have been proposed during the past two decades that optimize the sampling positions aiming to reduce the logical error. This paper focuses on minimisation of the logical error using a sampling design known as adaptive sampling.

Adaptive sampling has been used in dimensional metrology for many years, particularly when using coordinate measuring machines (CMMs). Although uniform, or regular, sampling is the mainstream for practical measurement, adaptive sampling has been shown to have an advantage in terms of efficiency. The origin of the methodology derives from allocation of dense sampling in key regions such as high curvature parts or high frequency parts, followed by use of less dense sampling in other regions. In this way, the measuring efficiency can be improved without significant deterioration in accuracy. Adaptive sampling is a sampling design which can redirect its sampling effort in response to the observed values [2].

Generally two main categories of the sampling design can be formed: these are CAD model based methods and non-CAD model based methods. The former specifies the sampling positions based on a given nominal CAD model by analyzing its local surface properties such as the mean curvature. Most previous work has developed sampling methods based on CAD models [3-6]. For example, two novel algorithms known as automatic sampling and surface curvature-based sampling were proposed by Elkott [3] for a given freeform model. In the work of Shih et al [5], three main sampling design methods – direct sampling, indirect sampling and local adjustment sampling – have been developed for general CAD model measurements. These methods have been shown to be advantageous in terms of sample size saving or accuracy improvement for most general cases. However, CAD model based solutions do not consider unexpected defects produced in common manufacturing practice and the positioning error in clamping the given work-piece may cause a significant positioning error. In contrast, the non-CAD model based method has the ability to adjust its sampling points in real time [7, 8] Edgeworth and Wilhem [7] proposed a real-time adaptive sampling method based on surface normal measurement. This solution avoids the potential positioning error and is able to effectively pick up the necessary information to identify potential defects. However, its sampling design may not be as optimized as in the case of the CAD-model based method. Also, the result is sensitive to the initial conditions, such as the initial sample position. It is clear that none of these methods are perfect and selection of the method used depends on the specific work-piece, and the required measurement efficiency.

In surface metrology, adaptive sampling has not previously been introduced in practical measurement. In fact, most of the existing characterisation parameters of surface texture, such as Sq, Sz and Ssk, are derived from statistical theory. Consequently an unbiased estimation of the population based on the selected samples is expected [9]. The uniform sampling method ensures each unit of a population has the same selection probability and this method is widely accepted and used at the present time. Although a few sampling methods such as the spiral sampling [10] have been introduced, the emergence of structured surfaces provides a situation in which novel sampling methods will be required.

Structured surfaces [11] are those whose surface texture is dominated by a deterministic feature pattern [12]. They are often on the micrometre scale and anisotropic in different directions. For example, a vee-groove structure on a blazed grating usually has a micrometre scale in one direction but millimetre scale in the perpendicular direction. Measurement of this kind of surface needs special attention because of the following.

1. Evaluation of micro-scale dimensions and form error parameters are usually of more importance than the more standard surface texture parameters.
2. CMM are usually not able to probe or resolve such micro-scale surface structures. Extraction of the dimensional or form error parameters from the surface texture data is the general measurement requirement.
3. Most surface structures require both large measuring range and small resolution as it is common for sharp curvature changes to be observed in practical applications and features may have high aspect ratios.

The above requirements mean that efficient sampling methods need to be generated and adaptive sampling is a novel solution to this problem. Unlike uniform sampling [13], spiral sampling [10] and low-discrepancy sampling [14] - which has a fixed sampling design if some sampling conditions (e.g. the sample size) are specified - adaptive sampling determines its sampling positions by responding to the prior sampling observations thereby self-optimising. By analysing the effect of initial sampling, the method is able to predict the position of areas of high curvature. This means that dense sampling can be targeted in these areas. Using this technique in combination with proper reconstruction algorithms [15] the original continuous surface can be steadily rendered for visual presentation and post-process geometry analysis.

Considering stylus profilometers that have a mechanism allowing for raster-scanning of samples, an adaptive sampling method known as indirect sampling [5] is further developed in this research. This proposed method has been tested on some typical structured surfaces by numerical simulation. The results have shown promising performance by reduction of the sampling duration and minimisation of the sampling error.

2. Another section of your paper

The sequential section adaptive sampling method is developed based on the indirect sampling method [5]. Considering the raster scanning mechanism used in stylus profilometry, this methodology comprises a two-stage algorithm: (1) profile adaptive compression sampling and (2) areal adaptive sampling.

2.1. Profile adaptive compression sampling

The core of the profile adaptive compression sampling method is a compression algorithm that is proposed in order to prune out the unnecessary samples for a given uniform sampling result. The method does not aim to reduce the sampling duration; on the contrary, it reduces the sample size by maintaining necessary reconstruction accuracy. After this process, those key samples which have significant influence on minimizing the reconstruction error are retained. A simulation result of this method is shown in figure 1 and it can be seen that dense samples are retained near the high curvature regions. The method proceeds as follows

1. For a given surface profile, obtain its digital measuring result using the instrument-permitted dense sampling setting (blue line in figure 1).
2. Divide the digital profile at inflection points, if any, into several segments that are solely concave or convex.
3. For each segment, evaluate the approximation error (usually the residual error between the original profile and the approximated/interpolated profile) of each segment.
4. If the error exceeds an initially set threshold (usually a fraction of the initial error from step 3), an extra sampling point is inserted on the profile curve at the midpoint. Otherwise, stop.
5. For each subinterval formed by insertion of a new point, repeat steps 3 and 4 until the approximation error is smaller than the threshold value.

(a) (b)

Figure 1. Comparison of the results of (a) a uniform sampling and (b) a profile adaptive sampling.

2.2. Areal adaptive sampling

Since profile adaptive compression sampling is simply a compression process, the real efficiency of this method comes from the areal adaptive sampling method. For a given surface, the methodology searches the key sample positions in the direction perpendicular to the main measuring axis; and then implements the profile adaptive compression for all the main axis profile measurements at the key positions. The technique comprises five steps as follows and as illustrated in figure 2.

1. Randomly (or uniformly) select *N* (usually ten) profiles parallel to the main measuring axis (X-axis in figure 2).
2. Implement profile adaptive compression sampling for each profile. Thus the key positions can be found.
3. Re-sort all the pruned key samples in accordance with their positions along the measurement axis (Y-axis in figure 2).
4. Downsample the key samples list produced in Step 3 by the factor *N* to prune out samples that are too dense.
5. For the downsampled key sample positions, implement the profile adaptive compression sampling for each profile on the main axis direction.

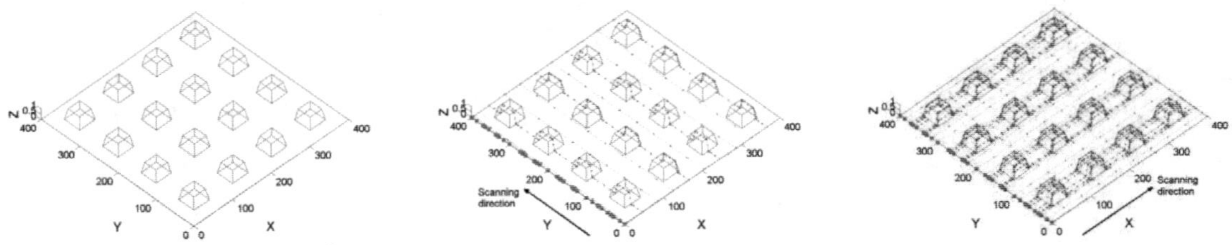

(a) (b) ©

Figure 2. Illustration of the areal adaptive sampling procedures (a) the original structured surface. (b) Implementation of ten profile adaptive compression sampling (dashed lines); and based on the pruned key sample positions (red dots), the downsampled key sample positions are selected (red squares). (c) Implementation of the profile adaptive compression sampling for each selected main axis (X-axis) profiles.

An improvement in measurement efficiency has been presented by the adaptive sampling result shown as the red sample points in figure2(c). Dense sampling intervals are arranged near the edges of the square step structures; while the low curved regions have a sparse allocation of samples. This method is thought to be useful for efficient measurement of linear and rectangularly tessellated structures for raster scanning profilometers. Using this method, the measuring size and duration can be effectively reduced and simultaneously reconstruction error, such as the residual root mean square (RMS) error and the error of the dimensional parameter evaluation, can be minimized. However, in the simulation results presented here, most of the general structured surfaces such as MEMS devices are also able to benefit from it.

3. Reconstruction and performance comparison

Since the sampling process transforms a continuous signal into its discrete form, reconstruction is necessary to make the representative continuous signal available from its discrete samples. Thus subsequent numerical analysis, such as FFT analysis and the extraction of surface texture parameters of the representative surface, can be implemented. Cubic spline interpolation has been introduced in profile reconstruction from non-uniform samples [16]; however, no robust reconstruction methods have been widely used in areal reconstruction of scattered sampling points. Considering that the scattered samples produced by the areal adaptive sampling usually have a parallelized distribution, a tensor product order 2 B-spline approximation solution [15] is used because of its computing

Journal of Physics: Conference Series **311** (2011) 012017 doi:10.1088/1742-6596/311/1/012017

efficiency and numerical stability. To quantitatively reveal the involved accuracy improvement, the uniform sampling and bilinear interpolation are used for comparison.

Table 1. Performance comparison of uniform sampling and the areal adaptive sampling.

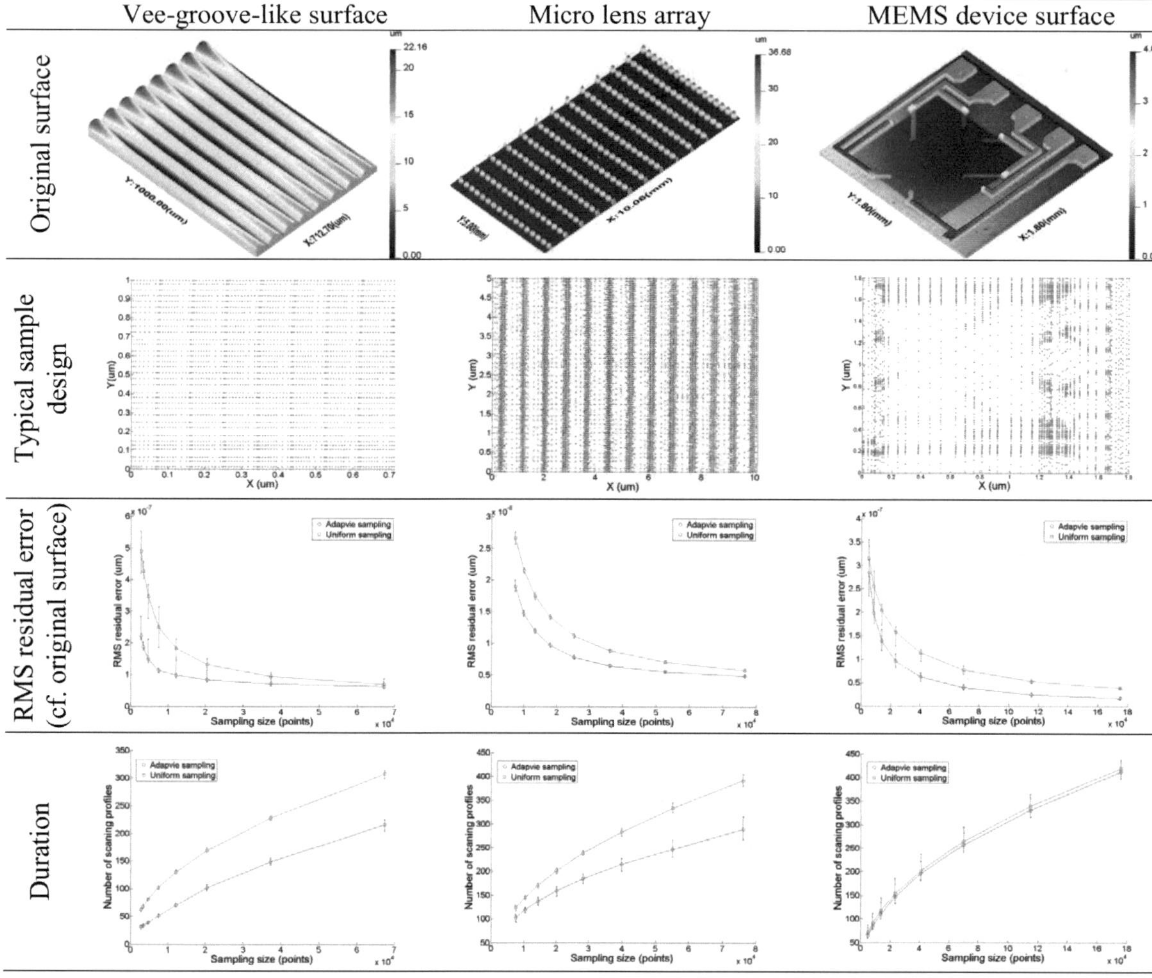

Two typical structured surfaces (vee-groove-like array structure and micro lens array structure) and a MEMS device surface are used as reference samples. They are presented in table 1 with an initial measurement using an extra dense sample setting. After that, areal adaptive sampling and uniform sampling are simulated. Three typical adaptive sample designs have been illustrated. By applying the reconstruction methods described earlier, the root mean square (RMS) residual errors compared to the original surfaces are analyzed. Simultaneously, the number of measuring profiles that are directly related to the measuring duration was computed. One hundred simulations were carried out with differing sample sizes so that repeatability could be quantified. By observing the results in table 1, aspects of performance improvement can be summarized.

1. This proposed adaptive sampling usually produces 20 % to 50 % lower RMS residual error than uniform sampling when using the same sample size.
2. When using the same sample size, the adaptive sampling usually uses 20 % to 50 % less measuring time than uniform sampling for the measurement of linear patterns and rectangular tessellations; while for general MEMS surfaces, no obvious difference can be observed.

3. At the same accuracy level, the adaptive sampling usually needs 40 % to 60 % less sample size than uniform sampling. It indicates that the adaptive sampling is able to achieve a 50 % to 80 % improvement in measuring duration saving when keeping the same accuracy level as uniform methods.
4. The two methods have nearly the same repeatability level while maintaining accuracy. However, on measurement duration estimation, the adaptive method generally presents a slightly reduced repeatability.

By applying this method to the case of raster scanning stylus profilometers, the sample size and measurement duration are able to be reduced effectively or the sampling accuracy can be improved significantly.

4. Conclusions

Based on the indirect sampling algorithm proposed by Shih et al [5], a novel areal adaptive sampling method has been developed. Applying this method to the case of raster scanning mechanism-based profilometers, the measurement efficiency for structured surfaces is expected to be significantly improved. In this work through simulation, the adaptive method has shown evidence of performance improvement compared to the currently used uniform sampling methods, although the repeatability remains relatively unchanged. Using this method in the measurement of the majority of structured surfaces such as linear patterns, rectangularly tessellated patterns and general MEMS device surfaces, the sampling reconstruction accuracy can be significantly improved or the sample size can be reduced at a significant level. The practical measurement duration can also be reduced for linear patterns and rectangular tessellations. Considering that current areal measurement duration for scanning system can often be hours, the proposed method is able to considerably increase measurement efficiency.

5. Acknowledgements

Jian Wang would like to thank the NPL UK for supporting this research work under the UK National Measurement System Programme for Engineering & Flow Measurement (2008-2011) and the University of Huddersfield under. X. Jiang would also like to gratefully acknowledge the Royal Society under a Wolfson Research Merit Award and the European Research Council under its programme ERC-2008-AdG 228117-Surfund.

References

[1] Shannon C E 1949 Communication In The Presence Of Noise *Proc. IRE* **37** 10-21.
[2] Thompson S K, Seber G A F and Okogbaa O G 1996 *Adaptive sampling*: Wiley New York
[3] ElKott D F and Veldhuis S C 2005 Isoparametric line sampling for the inspection planning of sculptured surfaces *Comput-Aided Des.* **37** 189-200
[4] Elkott D F, Elmaraghy H A and Elmaraghy W H 2002 Automatic sampling for CMM inspection planning of free-form surfaces *Int. J. Prod. Res.* **40** 2653 - 76
[5] Shih C S, Gerhardt L A, Chu W C-C, Lin C, Chang C-H, Wan C-H and Koong C-S 2008 Non-uniform surface sampling techniques for three-dimensional object inspection *Opt. Eng.* **47** 053606-15
[6] Cho M W and Kim K 1995 New inspection planning strategy for sculptured surfaces using coordinate measuring machine *Int. J. Prod. Res.* **33** 427
[7] Edgeworth R and Wilhelm R G 1999 Adaptive sampling for coordinate metrology *Precis. Eng.* **23** 144-54
[8] Hu J, Li Y, Wang Y and Cai J 2004 Adaptive sampling method for laser measuring free-form surface *Int. J. Adv. Manuf. Technol.* **24** 886-90
[9] Wieczorowski M 2001 Spiral sampling as a fast way of data acquisition in surface topography *Int. J. Mach. Tools Manuf.* **41** 2017-22
[10] Evans C J and Bryan J B 1999 "Structured", "Textured" or "Engineered" Surfaces *Ann. CIRP* **48**

541-56

[11] Jiang X, Scott P J, Whitehouse D J and Blunt L 2007 Paradigm shifts in surface metrology. Part II. The current shift. In: *Proc. R. Soc. A* **463** 2071-99

[12] Whitehouse D J and Phillips M J 1985 Sampling in a two-dimensional plane *J. Phys. A: Math. Theor.* **18** 2465-77

[13] Woo T C, Liang R, Hsieh C C and Lee N K 1995 Efficient sampling for surface measurements *J. Manuf. Syst.* **14** 345-54

[14] Barker R M, Cox M G, Forbes A B and Harris P M 2004 Discrete modelling and experimental data analysis *NPL Report*

[15] Brennan J K, Crampton A, Jiang X, Leach R and Harris P 2005 Approximation of surface texture profiles *J. Phys.: Conf. Ser.* **13** 264-7

Journal of Physics: Conference Series **311** (2011) 012018 doi:10.1088/1742-6596/311/1/012018

Surface topography analysis for dimensional quality control of replication at the micrometre scale

M Balcon*[1], F Marinello[1], G Tosello[2], S Carmignato[3] and E Savio[1]

[1] DIMEG, Dipartimento di Innovazione Meccanica e Gestionale, University of Padova, Via Venezia 1, 35131 Padova, Italy

[2] MEK, Department of Manufacturing Engineering and Management, Technical University of Denmark, Building 427S, Produktionstorvet, DK-2800 Kgs. Lyngby, Denmark

[3] DTG, Dipartimento di Tecnica e Gestione dei Sistemi Industriali, University of Padova, Stradella San Nicola 3, 36100 Vicenza, Italy

* Email: manuel.balcon@studenti.unipd.it

Abstract. Replication of geometrical features and surfaces are present at different production levels, from realization of moulds to final product. Geometrical features must be reproduced within specification limits, to ensure product functionality . In order to control the replication quality, mould and replica surfaces must be quantitatively analysed and compared. In the present work, reference simulated surfaces were considered and studied in order to evaluate the effectiveness and traceability of different analysis tools for replication quality control. Topographies were analysed simulating different surface mapping techniques, such as optical profilometry, scanning probe microscopy and scanning electron microscopy. Different strategies for surface analysis and comparison are proposed and discussed taking into account the instrument, the measuring range and the functionality of the surface.

1. Introduction

Replication is extensively used in micromanufacturing. Large majority of processes available for precision- and micro-material removal are relatively slow: creating and replicating a master is often the only profitable way for mass production. Another aspect of replication processes is connected to material selection: manufacturing technologies exploit their best performance only when applied to specific materials (e.g. electro discharge machining is not suited for polymer machining, milling is difficultly applied to brittle materials, or diamond cutting is not adequate for iron alloys). In all of these cases, making the workpiece using a first material optimized for the specific chosen technique and then replicating the geometry with a second material which better answers to the workpiece functionality requirements is a way to optimize the manufacturing process.

Hence, replication processes are involved at different levels of the production chains, from manufacturing to final products. The main drawback for implementation of replication in a process chain is the unavoidable loss of accuracy which is intrinsic to each replication step. Indeed, in any replication, there is loss of information during the shape transfer from the master to the copy [1]. Usually the geometry is reproduced only within a certain level of accuracy, and surfaces topography is not correctly replicated. The problem is evidently amplified whenever more than one replication step is introduced within the process chain. Therefore, in order to optimize the quality of the manufacturing process chain it is important to properly monitor replication steps: minimizing the difference between the master and copies allows assuring that the final product conforms to quality specifications [2,3].

In the present paper a study on the quality control of replicated surfaces is presented. The work is based on simulated data sets, resembling different surfaces, and in particular:

- master data sets resembling a surface with cylindrical and rectangular micrometric features (here on referred to as *structured surfaces*)
- master data sets resembling a generally planar surface with roughness and/or waviness contribution presenting no systematic distribution (here on referred to as *unstructured surfaces*)
- data sets of replicas simulated on both structured and unstructured surfaces (here on referred to as *replicas*)
- data sets resembling the output of different measuring instruments on both structured and unstructured surfaces or on replicas (here on referred to as *measurements*).

Techniques such as roughness analysis and cross correlation were applied to study the effect of the replications on the surface. Obtained results are discussed in order to identify instruments and analysis techniques that are sensitive to the replication process providing a reliable instrument for the quality control in the production of parts through replication processes.

2. Replication analysis at the microscale

Geometrical features must be reproduced accurately within tolerances. Surface characteristics and parameters, as for instance roughness or micro-geometry, influence in different ways functionality: tribological and mechanical behaviour, optical properties, appearance and tactile sensations transmitted by the surface. A correct replication eventually allows obtaining the desired surface properties in the final product. In order to control the replication quality, mould and replication surfaces must be analysed and compared. Different strategies for analysis and control of micro geometrical features surface replication are documented in literature, such as evaluation of roughness parameters, geometrical parameters, cross correlation, Fourier transform, etc. The choice of the analysis strategy depends on the product functionality that has to be controlled and, strictly related, there is the choice of the measurement instrument. Therefore the definition of an appropriate parameter to monitor a given replication process is the result of an optimal definition of the couple instrument-functionality.

Many metrological systems are available for dimensional and geometrical characterizations at the microscale. Scanning probe microscopes (SPM) consent to perform measurements with sub-nanometer accuracy providing three dimensional information about surfaces topography. Of particular interest are those instruments that can perform non-destructive measurements even on high aspect ratio structures such as 3D optical profilometers (OP) that produce fast non-contact measurements but are limited by the difficulty in achieving good reflection signal from surfaces that present high local slope or are reflective or transparent to the light source. Scanning electron microscopy (SEM) measurements allow resolutions on the nanometer scale, but must be used measuring conductive materials. Their major limitation is that, without additional measurement data and specific software tools, they perform only 2D measurements [4,5].

3. Generation of simulated data sets

Data sets of both structured and unstructured surfaces (as defined in section 1) were generated for preliminary investigations. Two different data sets were developed with lateral dimensions of 65.1 × 74.7 μm and maximum vertical size of 2 μm. The structured surface is characterized by the presence of well defined geometrical features: 5 positive and 5 negative rectangular features with constant depth (1 μm) and varying width (from 10 μm to 1 μm) and 6 circular features (3 positive and 3 negative, height 1 μm and diameter 5 μm).

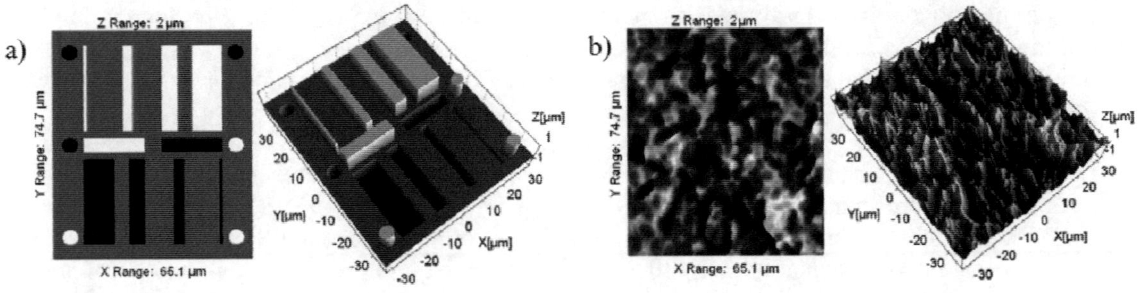

Figure 1. Data sets developed for the study: a) structured topography and b) unstructured topography.

Rectangular features are placed at a constant distance in order to realize a periodical structure. Periodical structures can be studied through a Fourier analysis. The width of the features is different in order to highlight the performance of different instruments when dealing with narrow and large structures (tip convolution for SPMs, reflection or shadowing for optical profilometers, and shadowing for SEMs). These kinds of structures allow amplifying replication artefacts as for instance the filling of channels during a polymer injection process.

The unstructured surface has an average roughness Sa = 698 nm. Such surface is a particularly interesting benchmark for SPMs and optical profilometers tests, providing a good base for tests with roughness parameters, cross correlation and Fourier transform.

Data sets were generated to simulate both master and replicas. Replicas (namely from *1* to *5*): simulated different degrees of replication accuracy, being *replica 1* a high accuracy replication and *replica 5* a low quality replication. In order to simulate an actual replication, the five grades of accuracy were artificially generated considering different levels of cavity filling, smoothing of edges and air inclusions. Additionally in the case of unstructured surface, simulated replicas where exhibiting also a lateral distortion, as it may happen when during the demoulding process, excessively high forces can cause a sample stretching distortion. Furthermore, three categories of measuring instruments were simulated: optical profilometers, scanning probe and scanning electron microscopes. Measurements with the different instruments were simulated including artefacts and distortions typical of the specific instruments. In order to simulate SPMs measurements, tip convolution, thermal drift and creep, noise due to vibrations and overshoots were considered. In OPs measurement simulations, void pixels (as for instance due to poor reflection, in high slope regions) and noise and spikes were included. SEMs images were simulated considering the two-dimensional characteristic of the instrument, and including shadowing, edge bloom and noise distortions.

All data sets were generated by means of a commercial software (SPIP™, by Image Metrology A/S, Denmark); replica and instrument simulations and cross correlation analysis were carried out taking advantage of specific routines developed and implemented in the software by one of the authors [6].

4. Surface characterization and comparison

Different data analysis approaches can be used in order to extrapolate quantitative information from the simulated topographies. Specific analyses have to be defined on the basis of the specific functionality relevant for the workpiece [7,8]. Most important analysis, considered within the present work, are:

1) Roughness parameters or areal statistical parameters, which give a description of the surface texture and allow comparing two or more surfaces. These parameters, together with the Abbott-Firestone curve, describe the distribution of peaks and valleys on the topography. Because of the statistical nature of the parameters, the use of this technique is normally helpful when describing an unstructured surface, but it is less useful whenever well-defined isolated geometrical features are

present on the surface. In this work the parameters considered were Sa, Sq, Sv, Sp, Spk, Sk, Svk as defined in ISO/DIS 25178-2 [8,9].

2) Volume related parameters: allow analysing the cavity filling of a replication process studying the volume of the cavities and the related volume of the replicated features.

3) Cross correlation index, applied to compare pairs of data sets. This index indicates how much two surfaces are overlapping, but it does not allow identifying where differences between master and replication surfaces are actually located [10].

4) Fourier related parameters. Such analysis gives information about the presence of periodic distributions and surface orientation. In particular, the comparison of the Fourier analysis between the master and the replication allows to control that the periodicity of the structure has been preserved during the replication process.

5. Results

Results related to unstructured surfaces are reported in figure 2 (first line). Values for replicas are reported as a percentage of the same value estimated in the master reference surface.

For an unstructured surface, the use of statistical roughness parameters is normally recommendable for analyses. These parameters are sensitive to the loss of surface quality from master to replica, but their variations result small. The main differences between master and replicas are in correspondence of highest peaks and deepest valleys, therefore only in a small portion of the total area: for these reasons they have only a negligible influence on parameters which are estimated in the whole areas, such as average roughness Sa. On the contrary, graphs show that cross correlation provide a useful index well describing distortions occurring from the master to the replica. Furthermore, roughness parameters are not sensitive to simulated horizontal distortions and present the same value obtained without distortions. This is a well known phenomenon, due to the fact that amplitude parameters are estimated only computing vertical elevation, and they are therefore insensitive to horizontal variations. On the other hand, cross correlation index and the wavelength of the surface calculated through Fourier analysis presented relevant variations (figure 2). As a general statement, the use of cross correlation method is recommended to describe replication accuracy when single or coupled horizontal and vertical distortions occur. Fourier analysis is also apt to detect and monitor lateral distortions, also in case of non periodical surface distributions, but on the other hand its behaviour is not influenced by vertical distortions, as clearly evidenced by the tests on structured surfaces.

The use of areal statistical parameters is normally not recommended whenever structured surfaces and isolated features have to be analysed. However, parameters related to the Abbot-Firestone Curve are representative of the bearing area regardless peaks and valleys distribution. Considering SPM simulations, the statistical parameters calculated in the master surface and in the replication surface, Spk, Sk and Svk are those that have shown the largest variations (see figure 2, second line). In particular Spk and Svk values decrease while decreasing the quality of the replicas. Conversely, Sk is increasing from replica 1 to replica 5. This is due to the fact that poor replication of the highest peaks and the lowest valleys causes an increase of the bearing surface. Hence, low performance features replication generates a more flat surface that can be monitored through an Svk and Spk decrease and an Sk increase. Other roughness parameters are less sensitive to the flattening of the surface. The variation of Sa, for example, is negligible. When calculating Sa, high local surface differences between master and replications (peaks and valleys), are considered distributed in the whole surface causing a small variation of the parameter even if the quality of replicas is low.

On the other hand, the same analysis performed on an optical profilometer didn't give the same results. This is mainly due to the fact that when measurements are carried on through an optical profilometer, surface regions locally overcoming maximum detectable slope (typically 30-40 degrees) are giving only a noise signal (or a so called "void pixels" portion, meaning a lack of data). Therefore in the case of optical profilometers, mainly flat regions are recognised: i.e. mainly top and bottom plateaux (when not shadowed) of the geometric structures and the mean plane. Therefore it is clear that, while a slight correlation with replication quality is still present when Spk and also Svk values are analysed, a lower correlation is noticed from Sk or Sa parameter analyses.

Interesting results were given by the cross-correlation index calculated between master and replicas. Significant variations were recognized with SPM (up to 48%) OP and SEM. The simulation however didn't consider the difficulty in relocating exactly the same surface portion (in master and in replicas), which is a factor particularly relevant and influent in cross-correlation computation.

Fourier analysis correctly gave only negligible variations (<1%, comparable with pixel resolution) on detected wavelengths of periodical structures for all of the three instruments (in figure 2, for readability reason, results are reported only in the case of SEM simulation).

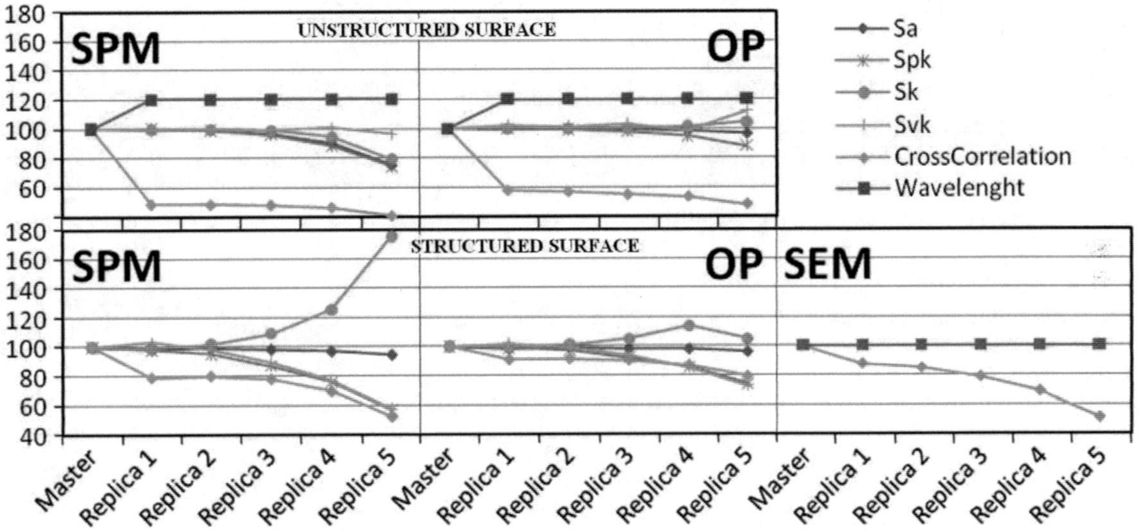

Figure 2. Parameters variations analysing different replicas in case of unstructured surface (first line) and structured surface (second line). Variations for different parameters (y-axis) are expressed as a percentage of the same parameter estimated in the master..

6. Conclusions

In this work artificial data sets were used to simulate replicated surfaces and to test their measurements through different instruments. The results were analysed in order to compare the quality of surface replication from the master to the final part. According with results presented above, roughness parameters related to the Abbot-Firestone curve can monitor the quality of a replica if the part presents isolated features extended along a direction perpendicular to the surface. Cross correlation index provides a description of the similarity between master and replications. When it is combined with a Fourier analysis can also highlight the presence of distortions during demoulding. However, the cross correlation is limited by the necessity to obtain two measurements of the same portion of the surface in the master and the replica.

References

[1] Alting L, Kimura F, Hansen H N and Bissacco G 2003 Micro Engineering *CIRP Annals* **52** 635-657

[2] Tosello G, Hansen H N, Marinello F and Gasparin S 2010 Replication and dimensional quality control of industrial nanoscale surfaces using calibrated AFM measurements and SEM image processing *CIRP Annals* **59** 563-568

[3] Tosello G, Hansen H N and Gasparin S 2009 Applications of dimensional micro metrology to

the product and process quality control in manufacturing of precision polymer micro components *CIRP Annals* **58** 467-472

[4] Hansen H N, Carneiro K, Haitjema H and De Chiffre L 2006 Dimensional Micro and Nano Metrology *CIRP Annals* **55** 721-743

[5] Leach R K 2010 *Fundamental Principles of Engineering Nanometrology* (Oxford: William Andrew Publishing)

[6] Marinello F 2007 *Atomic Force Microscopy in nanometrology: modeling and enhancement of the instrument PhD Dissertation* (Padova: University of Padova and Technical University of Denmark) http://paduaresearch.cab.unipd.it/1295

[7] Aris N F M and Cheng K 2008 Characterization of the surface functionality on precision machined engineering surfaces *Int. J. Adv. Manuf. Technol.* **38** 402–409

[8] Pawlus P, Galda L, Dzierwa A and Koszela W 2009 Abrasive wear resistance of textured steel rings *Wear* **267** 1873–1882

[9] ISO/FDIS 25178-2: Geometrical product specifications (GPS) - Surface texture: Areal/2: Terms, definitions and surface texture parameters

[10] Vorburger T V, Song J F, Chu W, Ma L, Bui S H, Zheng A and Renegar T B 2010 Applications of cross-correlation functions *Wear* (*In Press, Corrected proof* doi:10.1016/j.wear.2010.03.030)

Journal of Physics: Conference Series **311** (2011) 012019 doi:10.1088/1742-6596/311/1/012019

Effects of the cavity surface finishing on the polymer filling flow in micro injection moulding

G Lucchetta[*], F Marinello, G Tristo, M Balcon and M Berton

DIMEG, Dipartimento di Innovazione Meccanica e Gestionale, University of Padova, Via Venezia 1, 35131 Padova, Italy

[*] Email: giovanni.lucchetta@unipd.it

Abstract. In micro-injection moulding, dimension and shape of the cavity microscale factors, normally negligible in conventional injection moulding, play an important role. More specifically, cavity roughness, which in large scale moulding is relevant only for surface finishing and appearance, at the micro-scale can not only significantly change the cavity volume but also influence the polymer flow and the heat transfer between polymer melt and mould. In the present work the effect of the mould cavity finishing on the micro-injection moulding process was investigated. The phenomenon was analyzed using different channels for melt flow separation and taking advantage of localization of weld lines for differential characterization. To this end different micro-cavities with different surface finishing where generated by means of micro electro discharge machining (μEDM). Cavities surfaces where characterized integrating metrological information from different instruments: scanning probe microscopy (surface roughness), optical profilometry (roughness and volumes) and scanning electron microscopy (weld lines localization).

1. Introduction

Increasing demand for disposable products with micro- and nano-features is pushing research in many fields: from biomedics, to electronics, from micro mechanics, to optics. However, the development of new micro- and nano-scaled devices is highly dependent on the specific manufacturing processes that can produce micro parts in large quantities, reliably and economically. Micro-injection moulding (μ-IM) is a key technology for mass production of polymer micro-components, allowing replication of both miniaturized components and large components with miniaturized features. Currently, micro-injection moulding allows realization of components with sub-micron resolution and relatively high aspect ratios: however μ-IM performance significantly influenced by process parameters and by polymer characteristics [1, 2]. In a traditional injection moulding process, under regular injection velocity and cavity geometry, the melt polymer flow is laminar with a no-slip hydrodynamic boundary condition. On the other hand instabilities occur when geometry exhibits rapid changes, such as in the case of presence of gates or thin sections, especially if combined with high speed injection. The so called wall slip is a well established phenomenon for non-Newtonian fluids which can be ascribed to a disentanglement of the bulk chains when attached to the mould walls [3]. The effect is more problematic in injection moulding of small cavities. Indeed, when downscaling an injection moulding process, the rate between mould cavity surface and cavity volume increases, meaning that physical phenomena taking place at the polymer-mould interface increase their relevance and their influence on the whole process. Using computer tools to simulate and optimize a process is a well consolidated approach in traditional injection-moulding: but when sizes are downscaled, existing software packages are no longer sufficient to properly simulate the flow behaviour and cavity filling. Such inadequacy has to be ascribed to several limiting factors, but they all can be attributed to the fact that rheological properties of polymer used in current simulation packages are obtained from macroscopic scale measurements. Conversely, within micro scale geometries, surface tension and wall-slip conditions play a significant role, also depending on viscosity, heat transfer and temperature distribution, and roughness of cavity surface. More specifically, cavity roughness, which in large scale moulding is relevant only for surface finishing and appearance, at the micro-scale can not only significantly change the cavity volume but also influence the polymer flow and the heat transfer between polymer melt and mould. In the present work the effect of the mould cavity finishing on the micro-injection moulding process was investigated. The phenomenon was analyzed using different channels for melt flow separation and taking advantage of localization of weld lines for differential characterization.

2. Differential flow analysis

The method presented here for the first time is meant to analyze the specific effect of surface roughness on wall slip and on cavity filling. The method takes advantage of a double channel with a double link configuration, as reported in figure 1. The reference cavity can be divided in 4 main sections. In the first section (p_1 in figure 1) the channels are machined with the same width, roughness and volume. As a consequence, when the polymer melt flows meet in the second section (p_2) the weld line w_1 should be centred symmetrically between the two channels whenever the filling process is not influenced by any asymmetry, due for instance to a bad machining of the cavity, a different distribution of temperatures, etc. If this is the case, the position of the weld line in the first cross-link will be not centred: the displacement can then be used to verify any presence of asymmetries within the cavity and in case to compensate it. Then w_1 represent the "zero-position" for the following analysis carried out in the fourth section. The third section (p_3) is machined identically to the first one except for the surface roughness, which is different between the two channels. Any flow difference, to be ascribed to different wall-slip effects between the two channels, will be then quantitatively monitored in the fourth section (p_4) as a relative displacement on y direction of the second weld line w_2 with respect to the first one w_1.

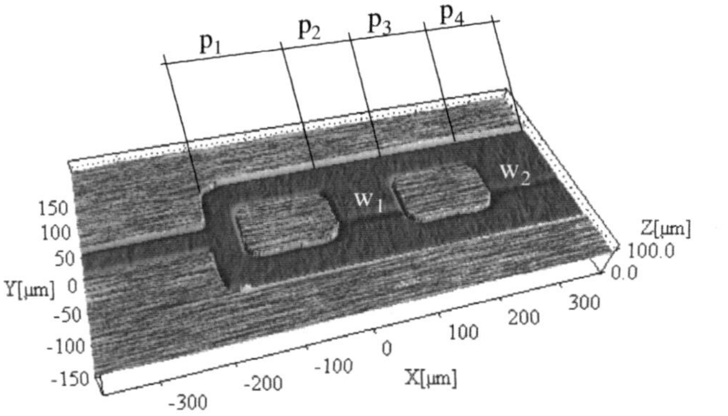

Figure 1. The newly design cavity filled with polymer with weld lines evidencing different wall slip behaviour.

2.1. Cavity manufacturing

To investigate the possibility of using the weld lines as flow markers to characterize the wall slip phenomenon during micro injection moulding, a micro cavity was designed and manufactured by micro electrical discharge machining (µ-EDM, Sarix SX 200). Thin channels, with width ranging from 40 µm to 200 µm and depth ranging 50 µm to 100 µm where produced with a central intersection (see figure 1 and figure 2a). Additionally, different machining processes were implemented in order to

Journal of Physics: Conference Series **311** (2011) 012019 doi:10.1088/1742-6596/311/1/012019

generate channels with a different surface finishing, with average surface roughness Sa comprised between 0.5 and 8 μm. In figure 2b an example is reported, where a reduction of applied voltage by 23% an increase of applied current by 30% and a reduction of energy by 50% allowed a reduction of average surface roughness Sa by almost 70%.

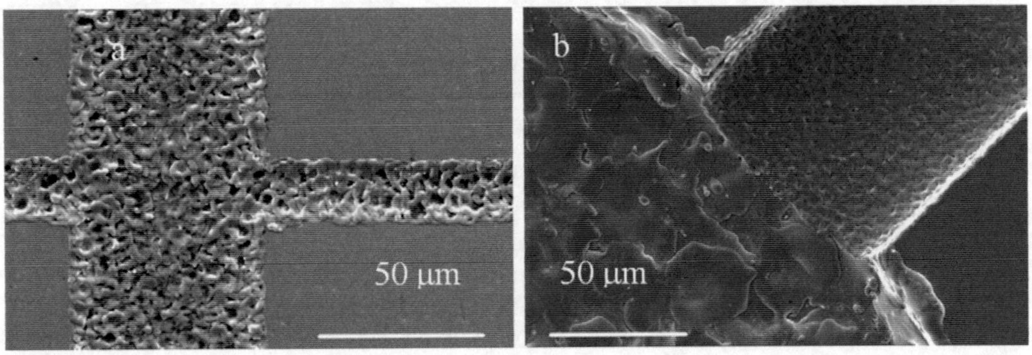

Figure 2. a) Micro machined channels, achieved by μ-EDM. b) Different surface finishing in terms of average surface roughness can be produced modulating μ-EDM crater average size, by modifying discharge parameters: large craters were achieved with a width of 5 μs, a frequency of 130 kHz, voltage 130 V, gap 65 V and relative energy 206; small craters with a width of 6 μs, a frequency of 100 kHz, voltage 100 V, gap 80 V and relative energy 105.

2.2. Cavity filling and polymer flow characterization

In order to evaluate the influence of surface finishing on wall slip, the cavity was filled at different injection speed using a micro injection moulding machine (Battenfled MicroPower 15). In parallel, a numerical simulation of the filling phase was conducted and the slip velocity was identified by inverse analysis, fitting numerical and experimental weld line position results. For the present work in an optical profilometer was implemented, allowing for 3D reconstruction of surface topography. This was interesting for our study in order to monitor at the same time not only the weld line position, the depth of the weld-line itself (which is a parameter of the quality of the injection moulding operation) and also the height of the polymer part with respect to the cavity height (which indicate the filling degree of the cavity before curing of the polymer flow).

3. Conclusions

The present paper reports a new method for measuring the effect of surface finishing on wall slip in micro injection moulding. By comparing numerical and experimental results regarding the position of the weld lines, it was possible to model the wall slip phenomenon in dependence of cavity surface finishing and geometry, melt speed and temperature and material properties.

References

[1] Tosello G, Gava A, Hansen H N, Lucchetta G and Marinello F 2009 Characterization and analysis of weld lines on micro injection moulded parts using Atomic Force Microscope *Wear* **266** 534-538

[2] Zhao J, Lu X, Lin M S, Chen G, Liu S L and Yong M S 2006 Effects of rheological properties of polymer blends on micro mold filling behaviour *Materials research innovations* **10** 111-112

[3] Sha B, Dimov S, Griffiths C and Packianather M S 2007 Micro-injection moulding: Factors affecting the achievable aspect ratios *Int. J. Adv. Manuf. Technol.* **33** 147–156

Journal of Physics: Conference Series **311** (2011) 012020 doi:10.1088/1742-6596/311/1/012020

Wear analysis through surface relocation

M Balcon*[1], F Marinello[1], S Carmignato[2] and E Savio[1]

[1] DIMEG, Dipartimento di Innovazione Meccanica e Gestionale, University of Padova, Via Venezia 1, 35131 Padova, Italy
[2] DTG, Dipartimento di Tecnica e Gestione dei Sistemi Industriali, University of Padova, Stradella San Nicola 3, 36100 Vicenza, Italy

* Email: manuel.balcon@studenti.unipd.it

Abstract. The present paper reports a new method for relocating three-dimensional surface topographies accurately, based on comparison between different displacement vectors and allowing compensation of lateral and rotational misalignments. Relocation is the main step for allowing exploitation of traceable quantification of surface modification, due to wear, etching or friction. The performance of the relocation routine was tested on a reference sample, featuring an array of peaks with a constant lattice of 2.12 μm.

1. Introduction

Characterization of surfaces and of surface evolution is a fundamental topic in precision engineering: scanning probe microscopy, as well as optical techniques can reproduce three dimensional topographies with very high accuracy [1]. On the other hand, inadequate stages cannot provide enough accuracy, and larger areas must be scanned in order to compensate errors occurring during positioning process.

Relocation post-processing operations are particularly valuable and necessary when comparison of corresponding areas is needed: this is the case of micro-engineered surfaces, replica analysis and fine surface evolution control. Finding exactly the same area is fundamental in order to properly characterize modifications and variation occurring with time between correlated surfaces.

Some work has already been proposed for relocation purposes, involving simple instrument relocation [2, 3], integration of reference grids [4], or use of simplex approach, based on reflection, expansion and contraction operations [5]. The present paper reports a new method for relocating three-dimensional surface topographies accurately, based on comparison between different displacement vectors and allowing compensation of lateral and rotational dislocations. Relocation eventually allows traceable quantification of wear volumes. Tests are reported on virtual images and actual surfaces, undergone to surface modification and subsequent estimation of worn volumes with the proposed technique.

2. Wear analysis

The developed method for analysis of surface evolution can be described according to the algorithm reported in figure 1. Firstly, topography measurements performed on the surface before and after modification (due for instance to wear, chemical etching or abrasive friction) are compared through the multiple cross-correlation relocation method, in order to recognise and compensate relative

Journal of Physics: Conference Series **311** (2011) 012020 doi:10.1088/1742-6596/311/1/012020

translational and rotational misalignment. Once the corresponding surface portions are recognised, the difference is calculated operating a point by point subtraction: at this stage, if necessary, a resampling can be performed in order to provide coherent data sets. At the last step the total difference is measured by means of root mean square parameter (Sq parameter in roughness analysis) or by analysing the residual volume.

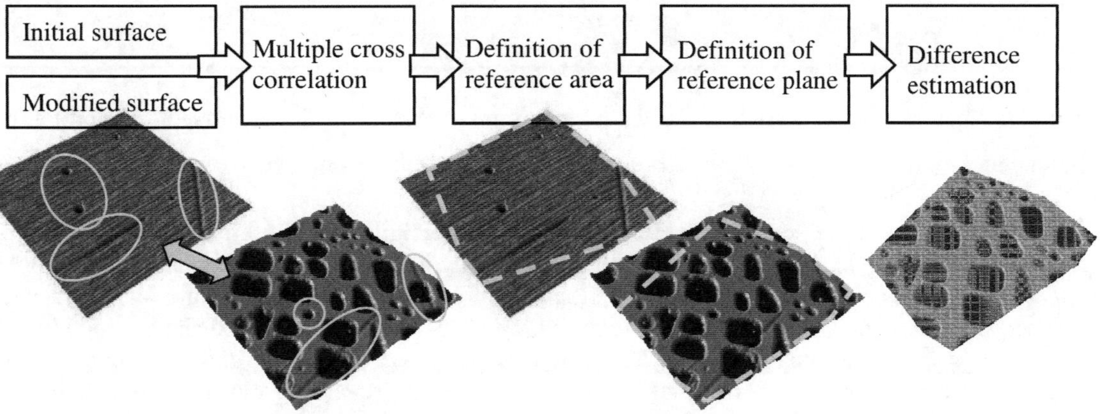

Figure 1. Algorithm and graphical representation for the estimation of surface variations.

3. Relocation

The developed relocation method is minded for couples of topographies, reproducing at least some portion of surface in common. Thanks to its simple architecture, results are achieved in a few seconds, even when heavy data file topographies are to be processed. The sequence of operations performed by the implemented relocation algorithm is described in the following; for simplicity, the topographies will be here on referred to as initial and modified surface.

1) Four areas are selected in the initial surface and four are correspondingly localized in the modified surface. Those are resembled by squares in figure 2.

2) Horizontal and vertical single pixel shifts are given to comparison areas in the modified surface: after each shift, a cross correlation function is calculated [6]:

$$cc = \frac{\sum_{x,y} [(z_r - \overline{z_r}) \cdot (z_c - \overline{z_c})]_{(x,y)}}{\sqrt{\left[\sum_{x,y} (z_r - \overline{z_r})^2_{(x,y)}\right] \cdot \left[\sum_{x,y} (z_c - \overline{z_c})^2_{(x,y)}\right]}} \tag{1}$$

where z_r and z_c represent the local surface height at the horizontal coordinate x, y in each of the four areas, respectively for the initial and for the modified surface; $\overline{z}_r$ and $\overline{z}_c$ are the mean values estimated in each of the four areas respectively in the initial and in the modified surface.

Best matching positions (blue squares in figure 2) are found by maximizing (1).

3) After best matching position has been found for each comparison area, displacement vectors are evaluated. In figure 2, vectors $\vec{v}_1$, $\vec{v}_2$, $\vec{v}_3$ and $\vec{v}_4$ are symbolized by black arrows.

4) Vectors are eventually processed and operated in order to find average translational (red dotted arrows in figure 2) and rotational misalignment (green arrows in figure 2).

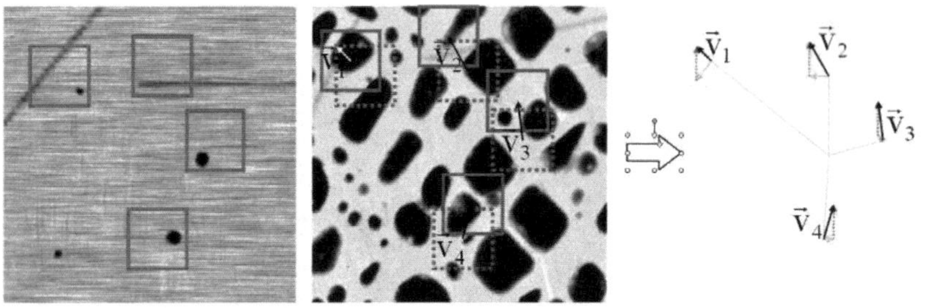

Figure 2. Vectors are processed and divided into translational ($\vec{v}_t$, dotted arrows) and rotational components ($\vec{v}_r$, green arrows); red and blue squares represent the detected corresponding areas.

The automatic routine was implemented into commercially available software for surface topographies analysis [7]. Investigations were performed by applying the developed software tool to surface topographies couples, reproducing a finely controlled rototranslational motion. Tests were performed on a silicon reference standard featuring an array of peaks (inverted tips) collated at the corners of a square grid, with a lattice of 2.12 µm (see figure 3a). This was eventually compared with the automatic routine output. In the figure below, main results regarding evaluated translation and rotation vectors are presented (figure 3b and 3c).

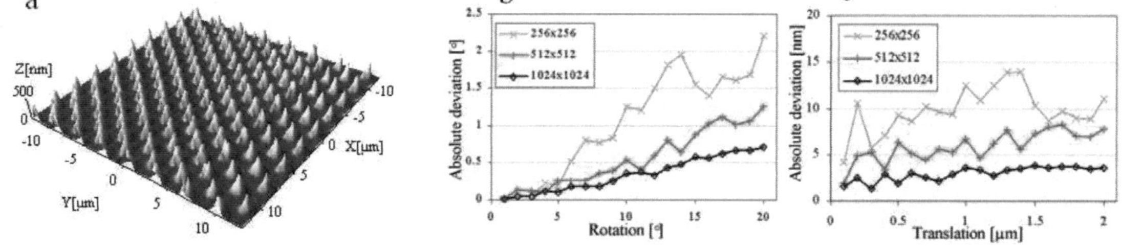

Figure 3. a) Reference sample used for testing. b) Absolute deviation between actual and evaluated rotation and (c) translation.

4. Reference area

Once the relative misalignment between the initial and the modified surface topography is compensated, surfaces portions present in both topographies are isolated while the rest is cut out. Thereafter, surface topographies are resampled, in order to have coherent data sets, with the same resolution and the same points relative positioning.

In order to verify where surface modification took place, corresponding areas have to be analysed again. The best strategy for recognising surface modification is to locally analyse cross correlation between corresponding surface portions. Positions where cross-correlation reduction exceeds measurement repeatability are recognised as "*modified positions*" (red shadowed areas in figure 1). Conversely, where surface topography modification is less than measurement repeatability (including noise and resolution) are recognized as "*unmodified positions*" (light blue shadowed areas in figure 1).

Unmodified positions are then used to define a reference plane for compensation of installation slope and surface flattening. On the other hand, once segmented *modified positions* are used in order to compute surface modifications.

Definition of unmodified positions is fundamental in order to define vertical relative positioning of corresponding surface portions. Such vertical alignment is indeed responsible for definition of the *zero* position, needed for measurement of volume or analysis of root mean square deviations. In figure 4 volume (represented by the blue region) is highlighted in the case of two different vertical positions for the reference *zero* line (dotted line): it is evident how such variation exponentially modifies volume definition.

Figure 4. Absolute deviation between actual and evaluated rotation (a) and translation (b).

From our experience a reliable definition of vertical relative position can be found when at least 25% of *unmodified* surface regions can be detected and used for comparison. To this end a key role is played by the accuracy of the measuring system. Indeed the lower is the repeatability of the measuring system, the lower will be the ability in operating a precise segmentation, from *modified* and *unmodified* surface portions.

5. Discussion and conclusions

The present paper reports a new method for relocating three-dimensional surface topographies accurately, based on comparison between different displacement vectors and allowing compensation of lateral and rotational misalignments. The automatic routine results are here summarized:

1) The developed method is based on an approximated algorithm: comparison areas are horizontally and vertically shifted in order to find local motions. In case of big rotations, rotational shifts should have to be taken in account: therefore the method is much incorrect as well as bigger rotations have occurred (figure 3b). This can be considered quite a minor problem, since stages can normally ensure an angle control with accuracy abundantly smaller than 5°.

2) Routine output is influenced by surface topographies resolutions: in fact higher deviations were detected whenever lower resolution topographies were analyzed (figure 3b and 3c). This problem can be overcome by considering a resolution software increase: in this way sub pixel accuracy can be achieved.

3) The developed method is not considering and evaluating deformations: for this reason these can lead to staggered results.

The developed method can be applied to all instruments for surface topography characterization: scanning probe, scanning electron, optical based microscopes, etc.

Main interests related with the new analysis tool are connected with exploitation of traceable quantification of surface modification, due to wear, etching or friction.

References

[1] Lonardo P M, Lucca D A and De Chiffre L 2001 Emerging trends in surface metrology *Ann. CIRP* **51/2** 701-723

[2] Pawlus P, Galda L, Dzierwa A, Koszela A 2009 Abrasive wear resistance of textured steel rings *Wear* **267** 1873-1882

[3] Cabanettes F, Claret-Tournier J, Mohlin J, Nilsson P H, Rosén B G and Xiao L 2009 The evolution of surface topography of injection cams *Wear* **266** 570-573

[4] Zhang H, Brown L, Blunt L and Barrans S 2009 Evaluation of fretting wear on the femoral stem surface *Int. Conf. on Bioengineering & Biomaterials*

[5] Condeco J, Christensen L H and Rosen B G 2001 Software relocation of 3D surface topography measurements *Int. J. Machine Tools and Manufacture* **41** 2095-2101

[6] Vorburger T V, Song J F, Chu W, Ma L, Bui S H, Zheng A and Renegar T B 2010 Applications of cross-correlation functions *Wear (In press, Corrected Proof* doi:10.1016/j.wear.2010.03.030)

[7] Scanning Probe Image Processor (SPIP^{TM}), developed by Image Metrology A/S, www.imagemet.com.

Journal of Physics: Conference Series **311** (2011) 012021 doi:10.1088/1742-6596/311/1/012021

3D surface topography of cylinder liner forecasting during plateau honing process

R Reizer[1], P Pawlus[2]
[1]University of Rzeszow, Institute of Technology, Rejtana 16A, 35-959 Rzeszow, POLAND
[2]Rzeszow University of Technology, W. Pola 2, 35-959 Rzeszow, POLAND

E-mail: rreizer@univ.rzeszow.pl

Abstract. Areal surface topographies after plateau honing process were measured. A correlation analysis of surface texture parameters was then carried out. As the results, the following parameters describing plateau honed cylinder 3D surface topography were selected: amplitude Sq, Sz, spatial: Str, Std, hybrid SΔq as well as functional: Spq, Svq and Smq. 3D surface topographies were modeled. The modeled surface topographies were correctly matched to measured ones in 77% of all analyzed cases. The plateau honing experiment was then carried out using an orthogonal selective research plan. Two machining parameters were input variables: coarse honing pressure p_v and plateau honing time t. Chosen cylinder liners texture parameters were output values. As the result of the experiment, regression equations connecting plateau honing process parameters p_v and t with recommended 3D surface topography parameters were obtained. Finally, cylinder liner surface topographies were predicted for various values of machining parameters. Proper matching accuracy of modeled to measured textures was assured in 67% of analyzed cases.

1. Introduction

The surface finish of cylinder bores is one of the most substantial factors influencing the friction, wear and lubrication of co-acting surfaces within the engine. Plateau honing process is commonly applied to receive the specific cylinder surface. The final topography is a result of two last steps in plateau honing process: coarse honing and plateau honing. Randomly generating surface roughness is simple and offers the following advantages: elimination of the hardware and software requirements, decrease of cost and time of experimental investigation. Usually non-Gaussian random surfaces with the desired skew and kurtosis were obtained by transforming the surface of normal ordinate distribution using the Johnson translatory system. Watson and Spedding [1] Uchidate et al. [2] as well as Gu and Huang used ARMA method [3], Hu and Tonder used FFT method [4], Bakolas used NCGM (Non-linear Conjugate Gradient) method [5]. Seong and Peterka [6] used FFT to generate a non-Gaussian pressure time series and so did Kumar and Stathopoulos [7]. Their models can be applied to simulation of non-Gaussian rough surfaces [8]. The method of two-process structure of plateau-honed cylinder liners surface simulation is the other possibility. It relies on the fact that fine part of the surface created during second process (plateau honing) has its characteristic amplitude distribution and a whole system represents transitional topography [9]. An idea of the proposed method (and its possible modifications) is imposition of random surface of Gaussian ordinate distribution (second process) on the base surface (first process).

2. Materials and methods

Work pieces from pearlitic grey cast iron with phosphorous eutectic of 220-260 HB material hardness were taken as samples. Plateau honing process was done using Gehring honing machine of type SZS150.M with NC II control. The HON 15 was the cooling fluid. Honing stones with ceramic binder were applied; for coarse honing: 02 8x10x150 5SGG 150 N10 VCA S5 but for plateau honing: 02 8x10x150 70C 500 15 V Ne. Honing angle was 53^0. As the results of initial investigations, the following variables in plateau honing process were selected: coarse honing pressure p_v and plateau honing time t. Plateau honing pressure p_p was constant during in research (0.8 MPa). Firstly a lot of samples (about 20) were subjected to plateau honing process. Then measurement of areal surface topography was conducted using Talyscan 150 stylus equipment. Measuring area was 2 mm x 2 mm, sampling interval was 5 μm in 2 orthogonal directions. The nominal radius of stylus tip was 2 micrometers. The measurement speed was 2 mm/s. After measurement, the form was removed by polynomial of 3rd level. No digital filtration was used. In order to eliminate highly correlated surface topography parameters, correlation analysis was carried out. As the result of correlation analysis, the set of parameters describing surface topography of plateau honed cylinder liners was selected. Then areal surface topographies after plateau honing were simulated. It was assumed that modeling accuracy was acceptable when the parameters of simulated surface topography were within confidence intervals for average parameters of machined (plateau honed) surfaces.

Secondly, the plateau honing experiment was carried out using orthogonal selective research design. Coarse honing pressure p_v (in the range: 4 – 6.4 MPa) and plateau honing time t (6.3-18.9 s) were input parameters. The experiment was also carried out in other research points than resulting from plan. As the result of experiment, the regression equations connecting plateau honing process parameters p_v and t with recommended surface texture parameters were obtained. Finally, surface profiles were predicted for various values of plateau honing parameters.

The following parameters were analyzed: arithmetical mean height of S-F surface Sa, root-mean-square height Sq, maximum pit height Sv, maximum peak height Sp, maximum height Sz, ten-point height S10z, the emptiness coefficient Sp/Sz, skewness Ssk, kurtosis Sku, surface section height difference corresponding to 20-80% of the material ratio SHtp, root-mean square gradient Sdq, arithmetic mean peak curvature Spc, developed interfacial area ratio Sdr, fractal dimension Sfd, texture aspect ratio Str, auto-correlation length Sal, density of peaks Spd and texture direction Sfd. In addition, the following areal parameters for stratified functional S-F surfaces were analyzed: Sk, Spk, Svk, Smr1 and Smr2 as well as Spq, Svq and Smq (ISO 25178-2).

3. Results and discussion

It was assumed that parameters were strongly correlated when absolute value of the linear correlation coefficient ρ was greater than 0.7. Amplitude parameters were strongly interrelated; especially Sa and Sq parameters. These statistical parameters were highly correlated with the following parameters: Sp, Ssk, Sp/Sz, SHtp, Sdq, Sdr, Spc, Spd, Svi, Sk, Svk and Smr1. However they were not statistically connected with parameters describing maximum surface height. Skewness was inverse proportional to kurtosis. Hybrid parameters Sdq, Spc and Sdr are also highly correlated. From among spatial parameters Str and Sal are independent of other parameters but mutually interrelated. The Std parameter was independent. The parameters Spq, Svq and Smq were mutually independent and also not correlated with the other parameters. Parameters from Rk family (Spk, Sk, Svq, Smr1 and Smr2) were mutually independent. On the basis of presented dependencies it was decided to describe surface topography of plateau honed cylinder liners by height parameters Sq and S10z, spatial parameters Str and Std, hybrid parameter Sdq and functional parameters Spq, Svq and Smq.

Table 1. Correlation coefficients for areal surface topography of cylinders after plateau honing (a)

Parameter	Sa	Sq	Sp	Sv	Sz	S10z	Ssk	Sku	Sp/Sz	SHtp	Sdq	Spc	Sdr	Spd	Str	Sal	Std	Sfd
Sa	1	0.942	0.472	-0.196	0.158	0.404	0.775	-0.582	0.723	0.979	0.987	0.956	0.988	-0.956	-0.085	-0.091	0.020	0.759
Sq	-	1	0.661	0.029	0.408	0.645	0.614	-0.453	0.754	0.865	0.933	0.835	0.914	-0.871	0.162	0.114	-0.045	0.553
Sp	-	-	1	0.377	0.823	0.884	0.163	-0.035	0.807	0.358	0.445	0.331	0.404	-0.405	0.490	0.444	-0.322	-0.038
Sv	-	-	-	1	0.837	0.591	-0.644	0.723	-0.182	-0.277	-0.221	-0.264	-0.247	0.238	0.525	0.484	-0.147	-0.573
Sz	-	-	-	-	1	0.885	-0.300	0.425	0.366	0.041	0.126	0.033	0.086	-0.092	0.611	0.560	-0.281	-0.375
S10z	-	-	-	-	-	1	-0.005	0.079	0.579	0.254	0.414	0.247	0.358	-0.330	0.467	0.424	-0.282	-0.180
Ssk	-	-	-	-	-	-	1	-0.908	0.626	0.805	0.774	0.797	0.763	-0.821	-0.319	-0.237	0.134	0.857
Sku	-	-	-	-	-	-	-	1	-0.477	-0.574	-0.613	-0.570	-0.580	0.639	0.444	0.322	-0.160	-0.718
Sp/Sz	-	-	-	-	-	-	-	-	1	0.665	0.711	0.632	0.684	-0.687	0.141	0.128	-0.236	0.430
SHtp	-	-	-	-	-	-	-	-	-	1	0.954	0.982	0.974	-0.951	-0.150	-0.152	0.063	0.818
Sdq	-	-	-	-	-	-	-	-	-	-	1	0.941	0.988	-0.961	-0.156	-0.169	0.011	0.766
Spc	-	-	-	-	-	-	-	-	-	-	-	1	0.960	-0.946	-0.200	-0.195	0.029	0.809
Sdr	-	-	-	-	-	-	-	-	-	-	-	-	1	-0.941	-0.150	-0.173	0.033	0.763
Spd	-	-	-	-	-	-	-	-	-	-	-	-	-	1	0.204	0.230	-0.058	-0.839
Str	-	-	-	-	-	-	-	-	-	-	-	-	-	-	1	0.877	-0.085	-0.457
Sal	-	-	-	-	-	-	-	-	-	-	-	-	-	-	-	1	-0.005	-0.477
Std	-	-	-	-	-	-	-	-	-	-	-	-	-	-	-	-	1	0.170
Sfd	-	-	-	-	-	-	-	-	-	-	-	-	-	-	-	-	-	1

(b)

Parameter	Sk	Spk	Svk	Smr1	Smr2	Spq	Svq	Smq
Sa	0.943	0.266	0.683	-0.627	0.160	0.280	-0.087	-0.540
Sq	0.819	0.505	0.827	-0.402	0.088	0.487	0.042	-0.407
Sp	0.292	0.692	0.664	0.101	-0.094	0.492	0.273	-0.029
Sv	-0.265	0.436	0.120	0.560	0.062	0.399	0.506	0.068
Sz	0.009	0.675	0.465	0.403	-0.018	0.533	0.472	0.024
S10z	0.188	0.753	0.734	0.167	-0.141	0.502	0.304	0.041
Ssk	0.772	0.037	0.395	-0.749	0.012	-0.009	-0.343	-0.315
Sku	-0.523	0.092	-0.423	0.698	0.165	0.095	0.374	0.037
Sp/Sz	0.588	0.465	0.653	-0.351	-0.140	0.282	0.009	-0.218
SHtp	0.986	0.166	0.522	-0.664	0.270	0.184	-0.147	-0.608
Sdq	0.907	0.248	0.715	-0.666	0.089	0.266	-0.082	-0.493
Spc	0.972	0.169	0.485	-0.650	0.286	0.167	-0.126	-0.586
Sdr	0.942	0.223	0.642	-0.638	0.185	0.223	-0.110	-0.536
Spd	-0.911	-0.179	-0.608	0.700	-0.111	-0.286	0.068	0.500
Str	-0.112	0.746	0.112	0.688	0.157	0.573	0.216	0.056
Sal	-0.118	0.664	0.137	0.533	0.113	0.349	0.154	0.137
Std	0.114	-0.116	-0.125	-0.145	0.051	0.048	-0.237	0.134
Sfd	0.795	-0.225	0.287	-0.846	0.054	0.068	-0.163	-0.488
Sk	1	0.160	0.415	-0.601	0.411	0.174	-0.172	-0.615
Spk	-	1	0.446	0.414	0.002	0.569	0.156	0.080
Svk	-	-	1	-0.334	-0.351	0.443	0.117	-0.102
Smr1	-	-	-	1	0.155	0.228	0.243	0.385
Smr2	-	-	-	-	1	-0.028	-0.176	-0.304
Spq	-	-	-	-	-	1	0.440	-0.025
Svq	-	-	-	-	-	-	1	0.309
Smq	-	-	-	-	-	-	-	1

Imposition method was used during creation of plateau honed cylinder surface. Firstly two crossed textures corresponding to surfaces created by coarse and plateau honing processes, were generated as follows. Two one-directional surfaces oriented in the measurement direction were obtained using the method developed by Wu [10]. These surfaces were rotated by angle Φ equaled to half of honing angle. Next initial one-directional surface was rotated by angle $180^0 - \Phi$ (see [11]). These two oriented surfaces were superimposed and as the results two crossed areal textures corresponding to plateau and valley part were obtained. Sq parameters of these textures were equal to desired values of Spq and Svq parameters of two-process surface. The distance between mean planes of these surfaces was determined in order to obtain desired value of Smq parameter. Then smaller ordinate values of

these two surface topography were selected. The special procedure was used in order to obtain values of other selected parameters describing plateau honed cylinder liner surface topography.
We obtained the following values of confidence intervals of measured surfaces: ±0.164 µm for Sq, ±1.624 µm for S10z, ±0.0115 for Sdq, ±0.0115 for Str, $\pm 0.046^0$ for Std, ±0.077 µm for Spq, ±0.621 µm for Svq and ±9.63% for Smq. The parameters Sq, Sdq, Spq, Svq and Smq of modeled surface topographies were correctly matched in 100%, the parameters Str, S10z and Std in 89%. The joint matching condition of all selected parameters was fulfilled in 77% of cases. The relative errors of parameter determination were: for Sq 0.77%, Sdq 5.08%, S10z 9.62%, Str 15.27, Std 1.1^0, Spq 7.09%, Svq 5.1% and Smq 4.25. The relative errors of other parameter determination were studied, too. They amounted: for Sa 9.49%, Sz 11.61%, Sp 12.34%, Sv 10.5%, Ssk 6.45%, Sku 15.32%, Spc 19.12%, Spd 22.525 and Sal 19.32%. Figure 1 presents example of measured and modelled surface. The parameter values are also given.

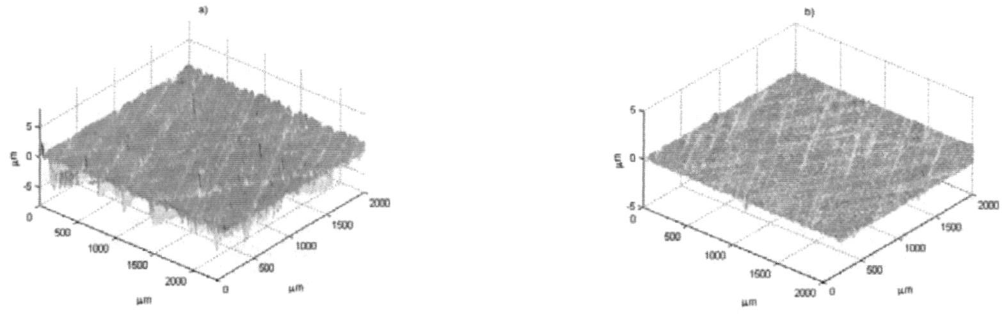

Sa = 0.55 µm, St = 7.75 µm, Sq = 0.87 µm, Sdq = 0.089, Spq = 0.221 µm, Svq = 2.214 µm, Smq = 76.3%, Sp = 2.5 µm, Sv = 5.25 µm, Ssk = -2.9, Sku = 13.45, S10z = 7.19 µm, Str = 0.047, Std = 26°, Spc = 0.01, Sal = 0.022 mm

Sa = 0.63 µm, St = 6.75 µm, Sq = 0.89 µm, Sdq = 0.088, Spq = 0.247 µm, Svq = 2.12 µm, Smq = 71.6%, Sp = 2.12 µm, Sv = 4.63 µm, SSk = -2.79, Sku = 13.36, S10z = 6.57 µm, Str = 0.048, Std = 26.5^o, Spc = 0.014, Sal = 0.021 mm

Figure 1. Isometric view of real (a) and modeled (b) cylinder lines surface topography after plateau honing

The plateau honing experiment was then carried out using orthogonal selective research plan (see Table 2). Standard deviations of parameters σ, obtained from 6 repetitions are also presented in Table 2. After removal of unsubstantial coefficient the following regression equations were obtained: Sq = 1.19 - 0.036 t, S10z = 11.69 - 0.22 t, Sdq = 0.25 - 0.013 t, Spq = 0.297 - 0.01 t, Svq = 0.44 + 0.094 t, Smq = 38.32 + 3.28 t, Str = 0.03 and Std = 26.5. It is evident that surface topography parameters depended only on plateau honing time t. The standard deviation of surface height Sq and of plateau surface part Spq, slope Sdq and the S10 z parameter was higher for smaller plateau honing time. However values of standard deviation of valley Svq and material ratio Smq were proportional to plateau honing time. The spatial parameters Str and Std were maintained on constant level. For the following machining parameters surface topography coefficients should obtain the following values:

- for p_v = 4.5 MPa, t = 15.3 s: Sq = 0.639 µm, S10z = 8.324 µm, SΔq = 0.0511, Spq = 0.144 µm, Svq = 1.878 µm, Smq = 88.504%, Str = 0.03, Std = 26.5^o,
- for p_v = 5.9 MPa, t = 9 s: Sq = 0.866 µm, Sz = 9.71 µm, SΔq = 0.133, Spq = 0.207 µm, Svq = 1.286 µm, Smq = 67.804%, Str = 0.03, Std = 26.5^o,
- for p_v = 6.4 MPa, t = 16.2 s: Sq = 0.607 µm, Sz = 8.126 µm, SΔq = 0.0394, Spq = 0.135 µm, Svq = 1.963 µm, Smq = 91.456%, Str = 0.03, Std = 26.5^o.

Table 2. The effect of plateau honing parameters on areal surface topography parameters

				Parameters						
p_v, MPa	**4**	**4**	**4**	**5.2**	**5.2**	**5.2**	**6.4**	**6.4**	**6.4**	**σ**
t, s	**18.9**	**12.6**	**6.3**	**18.9**	**12.6**	**6.3**	**18.9**	**12.6**	**6.3**	
Sq, µm	0.451	0.647	1.109	0.62	0.772	0.911	0.741	0.923	1.171	0.143
S10z, µm	6.193	7.771	9.24	7.501	8.693	10.201	7.547	9.503	10.107	1.412
Sdq,	0.052	0.073	0.126	0.064	0.086	0.101	0.091	0.096	0.107	0.01
Spq, µm	0.141	0.151	0.325	0.155	0.186	0.236	0.142	0.184	0.266	0.067
Svq, µm	2.608	2.215	1.555	1.944	1.778	1.555	3.58	1.723	1.492	0.54
Smq, %	90.027	79.140	34.860	79.697	76.713	53.177	83.300	79.290	40.933	8.38
Str	0.035	0.03	0.032	0.028	0.031	0.032	0.032	0.035	0.027	0.01
Std, ^{0}	26.5	26.5	26.5	26.5	26.5	26.3	26.5	26.5	26.5	0.4

Cylinder surface topographies with these parameters were modeled. We compared the results of simulation with parameters of measured surfaces for plateau honing parameters mentioned above. Matching accuracy of modelled to measured surfaces was fulfilled in 67% of cases. The parameters Svq and Smq of modeled surface topographies were correctly matched in 100%, the parameters Str, Sdq, Spq and Std in 78%, but the parameters Sq and S10z in 89% of all the analysed cases. Figures 2 and 3 present views of representative measured and modeled surfaces for presented above plateau honing process parameters.

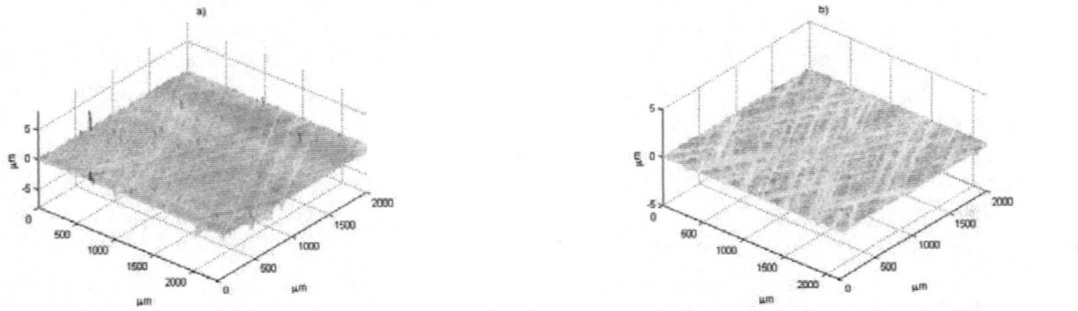

Sq = 0.616 µm, S10z = 8.18 µm, Sdq = 0.05, Spq = 0.15 µm, Svq = 1.52 µm, Smq = 94.1%, Str = 0.03, Std = 26°

Sq = 0.62 µm, S10z = 7.23 µm, Sdq = 0.04, Spq = 0.12 µm, Svq = 1.47 µm, Smq = 90.3%, Str = 0.09, Std = 26.5°

Figure 2. Isometric view of measured (a) and modeled (b) cylinder lines surface topography after plateau honing for the following machining parameters: p_v = 4.5 MPa, t = 15.3 s

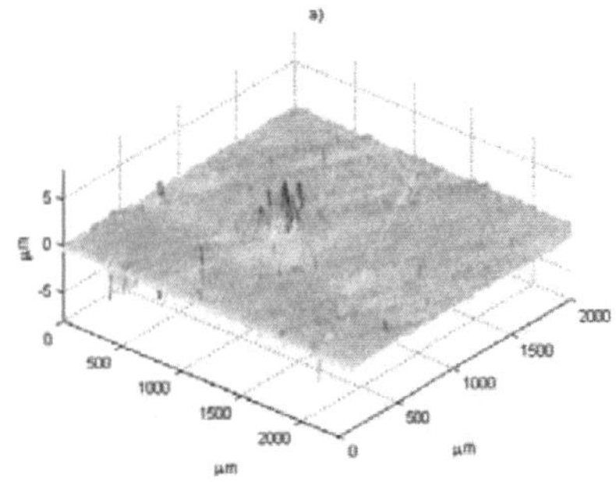

Sq = 0.49 μm, S10z = 7.22 μm, Sdq = 0.045, Spq = 0.14 μm, Svq = 1.8 μm, Smq = 94.1%, Str = 0.06, Std = 26°

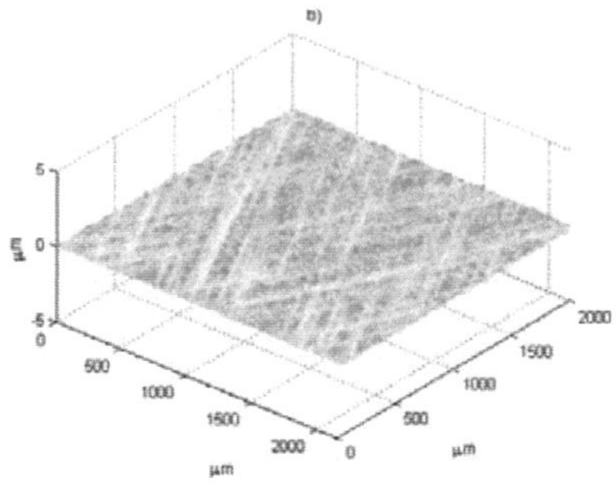

Sq = 0.51 μm, S10z = 6.22 μm, Sdq = 0.054, Spq = 0.17 μm, Svq = 1.91 μm, Smq = 91.5%, Str = 0.07, Std = 26.5°

Figure 3. Isometric view of measured (a) and modeled (b) cylinder lines surface topography after plateau honing for the following machining parameters: p_v = 6.4 MPa, t = 16.2 s

4. Conclusions

The following parameters were selected for description of plateau honed cylinder surface topography after plateau honing with stones of ceramic binder: Sq, S10z, Sdq, Spq, Svq, Smq, Str and Std. These parameters were chosen on the basis of their functional performance, the correlation analysis was also helpful in their selection. The modeled surface topographies were correctly matched to measured surface topographies in 77% of all the analyzed cases. The plateau honing experiment was conducted using an orthogonal selective research plan with two machining parameters being input variables: coarse honing pressure p_v and plateau honing time t. Selected previously cylinder liners texture parameters were output values. It was found that plateau honing time was the most important parameter affecting the following surface topography parameters: Sq, S10z, Sdq, Spq, Svq and Smq. The values of spatial parameters Str and Std were constant. The effect of coarse honing pressure on plateau honed surface topography parameters is negligible. It is possible to predict plateau honed cylinder surface profiles when machining parameters are known. Matching accuracy of anticipated to measured surfaces was fulfilled in 67% of all analysed cases.

5. References

[1] Watson W and Spedding T A 1982 *Wear* vol 83 pp 215-231
[2] Uchidate M, Shimizu T, Iwabuchi A, Yanagi K 2004 *Wear* vol 257 pp 1288-1295
[3] Gu X and Huang Y 1990 *Wear* vol 137 pp 275- 285
[4] Hu Y Z and Tonder K 1992 *Int. J. Mach. Tools Manufact.* vol 32 pp 82-90
[5] Bakolas V 2003 *Wear* vol 254 pp 546-554
[6] Seong S H and Peterka J A 1997 *Journal of Wind Engineering and Industrial Aerodynamics* vol 72 pp 95-105
[7] Kumar K S and Stathopoulos T 1999 *Engineering Structures* vol 21 pp 1086-1100
[8] Wu J –J 2004 *Tribology International* vol 37 pp 339-346
[9] Thomas T R 1999 *Rough Surfaces, Second Edition* (London: Imperial College Press)
[10] Wu J –J 2000 *Tribology International* vol 33 pp 47-58
[11] Reizer R 2011 *Wear* vol 271 pp 539–543

Journal of Physics: Conference Series **311** (2011) 012022 doi:10.1088/1742-6596/311/1/012022

Surface topography of cylindrical gear wheels after smoothing in abrasive mass, honing and shot peening

J Michalski[1], P Pawlus[1], W Żelasko[2]
[1]Rzeszow University of Technology, W Pola 2, 35-935 Rzeszów, POLAND
[2]Technical School Complex at Leżajsk, Mickiewicza 67, 37-300 Leżajsk, POLAND

E-mail: jmichals@prz.edu.pl

Abstract. The present paper presents the analysis of surface topography of gear teeth as the result of final machining processes. Teeth of multiple cylindrical gears shaped by grinding were smoothed in abrasive mass, honed or shot peened. The measurement of gears were made using coordinate measuring machine and 3D surface topography stylus instrument. The following deviations were studied; pitch deviation, total pitches deviations, variation of teeth thickness and deviation of gear radial run-out. Changes in teeth surface topography during machining process were determined. 3D surface topography parameters, surface directionality as well as areal autocorrelation and power spectral density functions were taken into consideration. As the results of the analysis, the best surface topography with regard to gear operational properties was recommended.

1. Introduction

The analysis of surface topography makes possible control of production process, prediction and optimization of machine elements functional properties [1]. It is widely known that method of gear tooth finishing and its surface topography highly influences gear fatigue life, its resistance to motion, vibration and noise [2-12]. Small value of power spectral density of teeth flanks surface topography assures little generated noise and vibration level of gear [11]. So it is desirable to optimize gear geometry. The problem of assessment of gear surface topography is of great practical importance.

Toothed wheels can be created by decrement, volumetric and incremental methods. It is required during contemporary gears production to carry out many tests, analyses and measurement. There are the following popular methods of toothed wheels finishing in soft and hard states: grinding, shaving, honing, super-finishing, lapping, shot peening, burnishing, electrochemical machining, chemical-abrasive treatment, magneto-chemical machining and others. The progress of shot peening [3, 4, 6], grinding [2, 7, 10, 13-15] and chemical-abrasive machining is the highest [9, 12].

2. The aims and scope of the investigations

Semi-finished products of toothed shafts from alloy low-carbon carburizing steel of AMS 6308 type were made by die forging method. Gear envelopes were quenched and tempered to hardness of 35–40 HRC. Module pitch of a gear in normal section was 1.90598 mm. Table 1 presents the characteristic of the analysed gears with their requirements.

Table 1. The characteristic values and requirements for rims of cylindrical gears

Specification toothed wheel rim	Value
1. Diametric pitch transfer plane, mm	2.03200
2. Diametric pitch normal plane, mm	1.90598
3. Normal pressure angle, °	25
4. Width toothed wheel rim, mm	34.23
5. Pitch diameter circle, mm	50.80000
6. Root diameter circle, mm	46.2534–46.1518
7. Number of teeth, -	25
8. Form diameter circle, mm	47.4472
9. Base diameter circle, mm	46.0404
10. Tooth fillet R, mm	min. 0,559
11. Form diameter roll angle, °, ′	32°11′
12. Based on chordal tooth thickness, mm	3.1877–3.2385
13. Reference data measurement over, mm, diameter wires 3,6576 mm	56.9264–56.8299

Journal of Physics: Conference Series **311** (2011) 012022 doi:10.1088/1742-6596/311/1/012022

14. Helix angle, °	20.31295°
15. External diameter circle, mm	55.5219

Gear teeth created by Sykes chiseling were subjected to carburizing, hardening, low-temperature treatment and tempering. Then grinding by shape method Kapp was executed with application of machine VAS 432 CNC with control system Optronic CNC-D140. Pitch feed of rough grinding was 0.05 mm per dual stroke of reciprocal motion, but longitudinal feed was 500 mm/min. Chemical-abrasive treatment of gears was made by smoothing in abrasive mass using machine made by Almco, model 29F-48A. This type of machining was done for both wheels after rough grinding by Kapp method as well as gears after rough and finish grinding. The first group (after rough grinding and smoothing) of toothed wheels was subjected to honing, but the second (after rough and finish grinding and smoothing) to shot peening. Five toothed shafts for each type of machining were tested.

Gear teeth honing was made using Fässler machine model D-250. Honing tool rim, of a teeth helical line slope 8° was equipped with abrasive material from aloxite with resinous binder 42A 180 R-5B. Shot peening was executed with steel balls S110 (of about 0.23 mm diameter) of Wheelabrator Allevard firm. Dry jet device using negative pressure method of Vapor Blast firm was used, VB 3576 type. The coefficient of surface coverage was 100%, shot peening intensity was F 13/17 A [3, 4, 6].

Stereometric shape of teeth sides was measured by coordinate measuring system CNC, model PNC 40 of Klingelnberg Söhne firm. Plunger finished by spherical surface of 1 mm radius was used. Table 2 presents the results of measurement.

Table 2. Accuracy parameters of side gear surfaces after machining (in brackets standard deviations)

Side of tooth deviations and run-out deviation	Finish grinding	Smoothing in abrasive mass	Honing	Shot peening
1. Single pitch deviations fpt, µm	2.3 (0.5)	1.6 (0.6)	1.4 (0.3)	1.5 (0.6)
2. Total cumulative pitch deviations Fp, µm	5.3 (0.8)	5.2 (1.3)	4.4 (1.1)	4.3 (1.1)
3. Pitch line run-out Fr, µm	9.1.1 (2.3)	4.7 (2.1)	2.2 (0.5)	4.8 (2.2)
4. Maximum variation of the chordal thickness Rs, µm	4.5 (0.5)	5.1 (2.0)	3.8 (0.6)	5.1 (2.1)

Tables 3 and 4 present the results of measurement of teeth sides. Talyscan 150 contact equipment of Taylor Precision firm was used. Square surfaces of side length 3.072 mm were measured. Sampling intervals in perpendicular directions were 6 µm. Form was removed by polynomial of 8-th degree. No digital filtration was used. 512 contours of tooth profile and 512 contours of tooth line were numerically separated. Definitions of parameters are presented in standards ISO 25178 and ISO 4287.

Table 3. Parameters of areal surface topography of tooth flank gears

Surface parameters	Grinding	Smoothing in abrasive mass	Honing	Shot peening
Sa, µm	1.30	0.26	0.11	0.33
Sq, µm	1.60	0.32	0.14	0.42
Sp, µm	6.09	2.40	3.88	3.49
Sz, µm	11.7	4.08	6.04	6.01
Ssk, -	-0.05	-0.24	1.35	-0.23
Sku, -	2.71	2.94	27.40	4.20
S10z, µm	10.40	3.46	4.41	4.77
Sv, µm	5.61	1.68	2.16	2.52
Str, -	0.0389	0.0117	0.0286	0.0421
Sal, mm	0.0440	0.0158	0.0295	0.0727
Std, °	4.0	4.5	3.0	1.5
Sfd, -	2.39	2.53	2.26	2.46
Spd, 1/mm^2	867	1843	2505	1617
Sdq, µm/µm	0.0820	0.0356	0.0175	0.0315
Spc, 1/µm	0.0120	0.0088	0.0049	0.0055
Sdr, %	0.3350	0.0634	0.0152	0.0495
Sk, µm	3.40	0.77	0.31	0.74
Spk, µm	1.21	0.24	0.13	0.31
Svk, µm	1.63	0.30	0.14	0.43
Smr1, %	9.9	7.4	10.4	11.2
Smr2, µm	81.8	86.3	88.4	87.2

Journal of Physics: Conference Series **311** (2011) 012022 doi:10.1088/1742-6596/311/1/012022

Table 4. Parameters of profile and line contours of side tooth after machining

Profile parameters	Contour of profile			Contour of line		
	Smoothing in abrasive mass	Honing	Shot peening	Smoothing in abrasive mass	Honing	Shot peening
Pa (μm)	0.26	0.11	0.30	0.26	0.11	0.30
Pq (μm)	0.32	0.14	0.38	0.62	0.30	1.61
Pp (μm)	1.26	1.00	1.19	1.26	1.00	1.19
Pt (μm)	2.86	1.71	2.54	2.86	1.71	2.54
Psk (-)	-0.24	0.04	-0.29	-0.24	0.04	-0.29
Pku (-)	2.90	4.61	2.86	2.90	4.61	2.86
Psm, (μm)	63	78	72	84	64	78
PΔa (°)	1.396	0.445	1.092	0.434	0.279	0.244
PΔq (°)	1.786	0.593	1.404	0.593	0.414	0.370

3. Results and discussion

Accuracy of side surfaces of toothed wheel is the highest after Fässler honing. It results from the values of deviations: fpt, Fp and Fr. Maximum deviation Rs as well as standard deviations of the analysed accuracy parameters are also the smallest after honing. The values of side surfaces of toothed wheels after grinding by Kapp method also decreased after smoothing in abrasive mass and shot peening. Profile contours on tooth height after grinding are not smooth in geometric sense. During smoothing in abrasive honing and during shot peening the flank tooth surface topography changes are small. However Fässler honing makes possible to obtain smooth shape of tooth contour on its height. The other important property of honing is the possibility of creation of required tooth shape and also its changes in comparison to that obtained previously in grinding process.

Roughness height of teeth surfaces significantly decreased as a result of smoothing in abrasive mass after earlier grinding. Not only roughness height decreased, but also fastest decay autocorrelation length Sal, root mean square slope Sdq and arithmetic mean peak curvature Spc. Parameters characterizing material ratio curve: core roughness depth Sk, reduced peak height Spk and reduced valley height Svk also decreased.

Honing of gear teeth after rough grinding by Kapp method and smoothing in abrasive mass by Almco method assures the smallest roughness height. The lack of deep valleys and presence of not numerous peaks is its characteristic feature. It is also characterized by the highest obtained peak density Spd. Small values of the following parameters prove its profitable properties: root mean slope Sdq, peak curvature Spc, core roughness depth Sk, reduced peak height Spk and reduced valley height Svk.

Shot peening of teeth sides previously ground by Kappa method, smoothed in abrasive mass by Almco method, in small degree caused increase in maximum surface height Sz. Average height parameters Sa and Sq increased significantly. Autocorrelation length Sal and texture parameter Str also much increased. However skewness Ssk, kurtosis Sku and angle Std decreased. Surface directionality changes are small. However changes of power spectral density are high.

Profitable functional features of side teeth surface after Fässler honing were confirmed also by small height of contours of profile and line. That machining method creates also large profile spacing, PSm parameter value proves it. The profitable features were confirmed by the smallest slope Sdq and the smallest peak curvature Spc. Similar conclusions may be drawn from the analysis of profile and line slope.

Anisotropy of tooth side is rather high after shot peening (previously and after rough and finish grinding by Kapp method and smoothing in abrasive mass) - see values of the Str parameter. All the analysed machining methods create tooth flank surface of random-deterministic (quasi-deterministic) type. Autocorrelation function and power spectra density function pointed to high surface directionality along tooth side. Power spectral density is characterized by comparatively high values for large wavelength after grinding by Kapp method and smoothing in abrasive mass by Almco method. This surface is crossed with small angle. Fässler honing assured the smallest value of power spectral density. Power spectrum was separated particularly for large wavelengths.

After the analysis of angular spectra of areal autocorrelation and power spectral density functions one can see that angular positions of mean value of angular spectra (non modal value) are measures of surface direction. Quasi-periodical structure of machined tooth sides, particularly of high correlation length in tooth line direction caused that it is impossible to obtain information about surface directionality on the basis of cross-correlation function. For all the analyzed machining methods areal autocorrelation function and power spectral density function present both surface directionality and characteristic of background. Similar assessment of surface directionality can be obtained on the basis of power spectral density and autocorrelation function, but the first of them seems to be better. Amplitude distributions after smoothing in abrasive mass by Almco method and shot peening by Vapor Blast method are similar to each other, but different distributions were achieved after honing by Fässler method.

Journal of Physics: Conference Series **311** (2011) 012022 doi:10.1088/1742-6596/311/1/012022

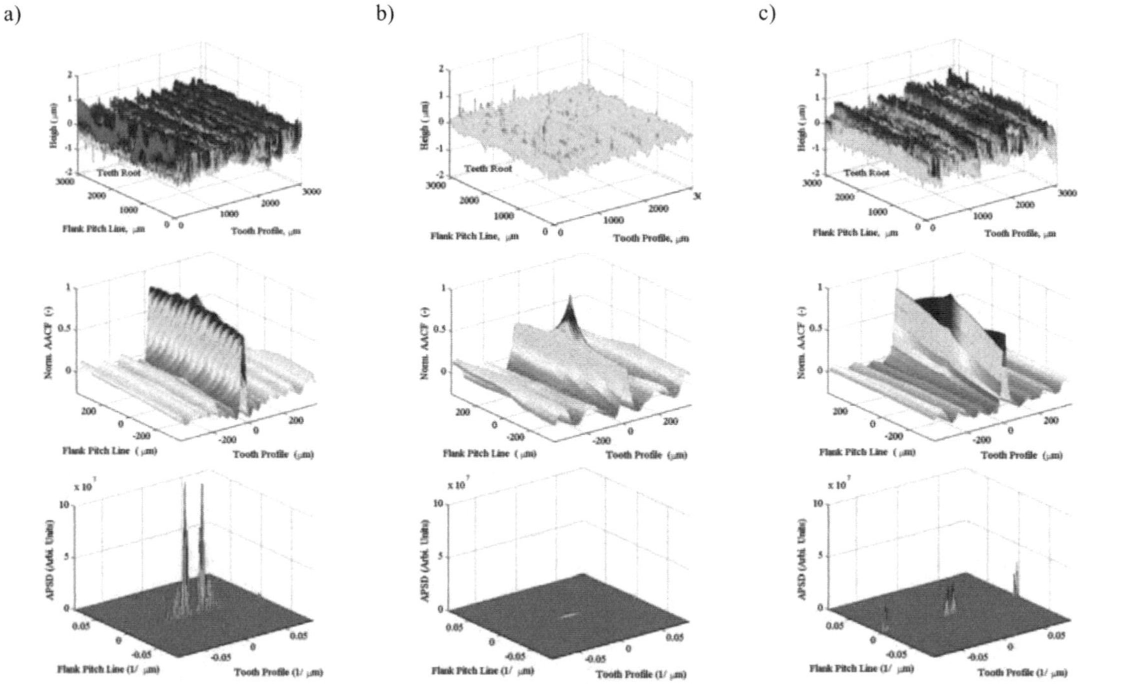

Figure 1. Isometric view, areal autocorrelation function AACF and power spectral density function for side of cylindrical gear tooth a) smoothed in abrasive mass, b) honed, c) shot-peened

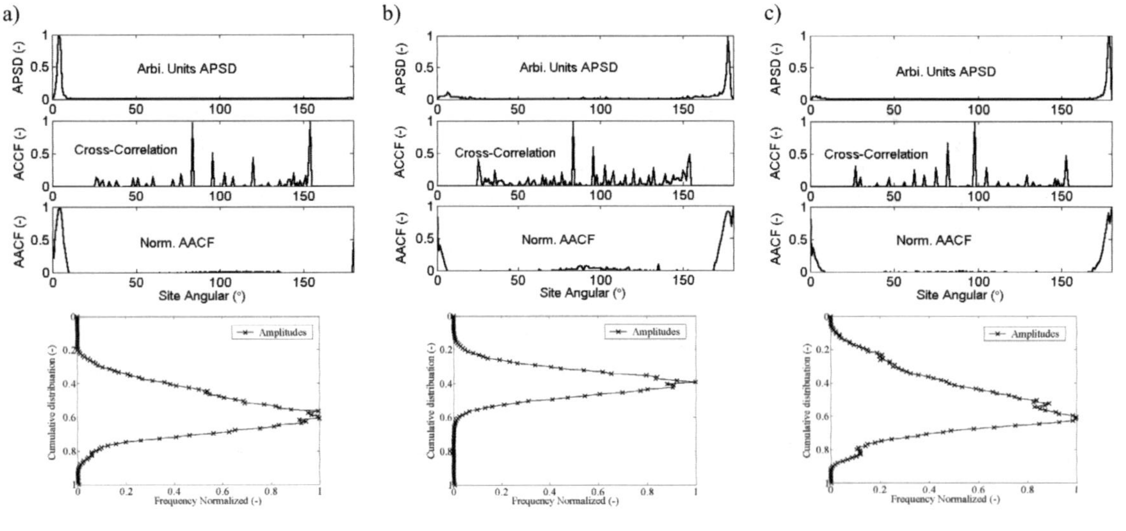

Figure 2. Angular spectrum of areal power spectra density (APSD) function , cross-correlation function (ACCF) and areal autocorrelation function (AACF) and amplitude ordinate distribution for side of cylindrical gear tooth a) smoothed in abrasive mass, b) honed, c) shot-peened

4. Concluding remarks

Surface of honed tooth side in hard state, after previous rough grinding by Kapp method and smoothing in abrasive mass by Almco method is characterized by the smallest values of kinematic accuracy deviations fpt, Fp, run-out Fr and chordal tooth thickness Rs. This surface was recommended for gear of small resistance to motion, vibration and noise. Small roughness height and low power spectral density prove about profitable properties. The possibility of precise and reproducible modification of tooth profile along its height and tooth line along its width is the other advantage of Fässler honing.

The results of application of smoothing in abrasive mass and shot peening are similar with regard to creation of amplitude ordinate distribution of tooth side surface topography; however honing by Fässler method caused better ordinate distribution characteristics.

The highest possibility of modification of geometrical features of tooth side was found for teeth shot peening by Vapor Blast method after earlier rough and finish grinding by Kapp method and smoothing in abrasive mass. Shot peening is contemporary applied to finishing of tooth sides after pretreatment by honing. Kapp grinding assures high production efficiency. Presented production process of shot peening by Vapor Blast method assures similar tooth profile and line accuracies to these created in previous machining process – finish grinding. Although roughness height after shot peening is much higher that after smoothing in abrasive mass, changes in power spectral density values are small.

Similar and proper assessment of surface directionality can be obtained on the basis of power spectral density and autocorrelation functions, but the first of them seems to be better. Surface of tooth flank side after machining by the analysed method is anisotropic, it is also strongly oriented along teeth lines.

References

[1] Goch G 2003 Gear metrology *CIRP.Manuf.Tech.* 52 659~695.
[2] Krantz T L Alanou M P Evans H P Snidle R W 2001 Surface fatigue lives of case-carburized gears with an improved surface finish *ASME.J.Tribol.* 123 709~716.
[3] Pariente I F Guagliano M 2009 Contact fatigue damage analysis of shot peened gears by means of X-ray measurements *Eng.Fail.Anal.* 16 964~971.
[4] Prasannavenkatesan R. Zhang J. McDowell D L Olson G B Jou H -J 2009 3D modelling of subsurface fatigue crack nucleation potency of primary inclusions in heat treated and shot peened martensitic gear steels *Int.J.Fatigue* 31 1176~1189.
[5] Starzhinskii V E Soliterman Yu L Goman A M Osipenko S A 2008 Forms of damage to gear wheels: Typology and recommendations on prevention *J.Friction.Wear* 29 340~353.
[6] Soyama H Macodiyo D O 2005 Fatigue strength improvement of gears using cavitation shotless peening *Tribo.Lett.* 18 181~184.
[7] Britton R D Elcoate C D Alanou M P Evans H P Snidle R W 2000 Effect of surface finish on gear tooth friction *ASME.J.Tribol.* 122 354~360.
[8] Wei B -Y Deng X -Z Fang Z -D 2007 Study on ultrasonic-assisted lapping of gears *Int.J.Mach.Tools.Manuf.* 47 2051~2056.
[9] Xiao Li Rosén B -G Amini N Nilsson Per H 2003 A study on the effect of surface topography on rough friction in roller contact *Wear* 254 1162~1169.
[10] Xiaoa Li Rosen B -G Amini N 2004 Surface lay effect on rough friction in roller contact *Wear* 257 1301~1307.
[11] Naser Amini Rosén B -G Westberg H 1998 Optimization of gear tooth surfaces *Int.J.Mach.Tools.Manuf.* 38 425~435.
[12] Oobayashi K Irie K. Honda F 2005 Producing gear teeth with high form accuracy and fine surface finish using water-lubricated chemical reactions *Tribol.Int.* 38 243~248.
[13] Brinksmeier E Giwerzew A 2005 Hard gear finishing viewed as a process of abrasive wear *Wear* 258 62~69.
[14] Karpuschewski B Knoche H -J Hipke M 2008 Gear finishing by abrasive processes *CIRP.Manuf.Tech.* 57 621~640.
[15] Marzenell C Tönshoff H K 2001 Properties of tooth surfaces due to gear honing with electroplated tools *Gear.Tech.* 18 43~49.

Problems of surface topography with oil pockets analysis

W. Grabon, P. Pawlus*, L. Galda, A. Dzierwa, P. Podulka

Rzeszow University of Technology, W. Pola 2 st., 35-935 Rzeszow, Poland

E-mail: ppawlus@prz.edu.pl

Abstract. The present paper presents method of description of two-process surfaces of random (load-bearing plateau structure) and deterministic (oil pockets) character on the basis of the probabilistic approach of material ratio curve. Examples of this method usage for textured surfaces are shown. The other application of material ratio curve for estimation of pit-area ratio of textured surfaces is discussed. Distortion of surface topography with oil pockets measurement by improper selection of areal reference plane is taken into consideration.

1. Introduction

Surfaces with a dominant deterministic feature pattern are termed "structured" or "textured" surfaces. Structured surfaces have numerous applications ranging from optics to automotive, from aerospace to biomedical and from micro-fluidics to power generation. The key feature that determines a structured surface is that its topography is not just an artifact of the process used to generate the surface, i.e. it has been engineered for a specific function. Thus, for a structured surface, typical parameters such as Ra or Rq do not adequately characterize its properties. Therefore, such surfaces are a challenge to manufacture and a challenge to measure. However, their function is by definition profoundly affected by their geometrical characteristics. Example includes surfaces to control the tribological characteristics of mating components. Method of this surface type creation is termed as surface texturing.

Recently, surface texturing attracted the attention of tribologists since it can be used to improve tribological properties of sliding elements without changing the materials and the lubricants. The dimples can serve either as a micro-hydrodynamic bearing in cases of full or mixed lubrication, a micro-reservoir for lubricant in cases of starved lubrication conditions or a micro-trap for wear debris in either lubricated or dry sliding [1, 2]. Various techniques can be employed for surface texturing including laser texturing, etching techniques and machining. It seems that laser surface texturing offers the most promising concept, because the laser is extremely fast and allows short processing times, is clean to the environment and provides excellent control of the shape and size of the texture, which allows realization of optimum designs [2]. A burnishing (embossing) technique is very promising in the creation of textured surfaces [3]. The applications of texturing include: cylinder liners [4, 5, 6], piston ring [7, 8], piston pin, disk brakes, thrust bearings, journal bearings [3], mechanical face seals, gas seals, hard disk sliders, magnetic tapes, machine tools guideways and other elements. The textured surface topography is functionally important in tribology, but it is notoriously difficult to characterize. Characterization of it entails analysis and identification of the departure from the nominal geometry of the basic repeat unit and analysis of the distribution and spacing of the repeat unit [9, 10]. The approach used to these surfaces is analyzing the surface autocorrelation function and then applying pattern analysis (segmentation) to the autocorrelation function. The segments around the origin of the autocorrelation function provide information on the average basic repeat unit and the relationship of the origin segment with its neighbouring elements of the tessellation pattern. The mean distances as well as the maximum, minimum and average depth, diameter of the holes and pit-area ratio were computed. The diameter is calculated as the best fit least square circle fitted to the segmentation boundary of the holes.

In this paper some problems of measurement and analysis of surface topography with oil pockets is discussed. The authors focused on the application of material ratio curve for determination of tribologically important surface topography parameters and on selection of reference plane for cylindrical surfaces with oil pockets.

2. Application of material ratio curve

Two scales of roughness can be found on textured surface: a relatively rough valley structure containing dimples and comparatively smooth finer plateau structure. When plateau surface part was created by honing or grinding or arose during wear it had random character, but valley structure, created by burnishing technique deterministic character. The random character of plateau part was confirmed by the analysis of the surface details free of dimples and by the fact, that the fine part of textured surfaces can be approximated by a straight line in material probability plot. Material probability curve is a representation of material ratio curve in which the surface material ratio is expressed as Gaussian probability in standard deviation values, plotted linearly on the horizontal axis. This scale is expressed in standard deviation s. For stratified (two-process) surfaces, composed of two Gaussian distributions, the material probability curve exhibits two linear regions. Ppq parameter is the slope of a linear regression performed through the plateau region, but Pvq – through the valley region. The intersection point on normal probability graph of abscissa Pmq defines the separation of plateau and base textures. These parameters were defined in ISO 13565-3 standard. Since this approach independently characterises the components of the multi-process texture, process control can be much more specific. However ISO 13565-3 standard is devoted only to stratified surfaces composed of two Gaussian distributions. Therefore for the analyzed textured surfaces of random-deterministic character Pq parameter (height standard deviation) of surface details free of dimples equalled to Ppq parameter of worn textured surface can be determined from the analysis of surface fragments between dimples. However this procedure is time consuming. One of the present authors developed the method of Pqp and Pmq parameters determination for this type of surface. This problem is of great practical importance because standard deviation of height in areas free of dimples can affect tribological properties of sliding pairs. The main problem is determination of transition point between random and deterministic regions. This point was determined by rotation of material probabilitty plot of ψ angle anticlockwise according to the following equation:

$$\begin{aligned} x &= x\cos\psi - y\sin\psi \\ y' &= x\sin\psi + y\cos\psi \end{aligned} \tag{1}$$

ψ angle is the slope of straight line passing by the first and the finishing point of the material ratio curve (see Figure 1). In rotated diagram C point of the highest ordinate was determined (see Figure 2). This point is treated as transition between deterministic and random regions. The regression line passing by points lied on point C left side was determined (see Figure 3). Slope of this straight line is equal to the Pqp parameter, but abscissa of C point is equal to transition Pmq parameter. Figure 4 shows example of the presented method application. Areal Spq or Smq parameters can be also determined in this way. After the analysis of many surfaces it was found that this method is useful.

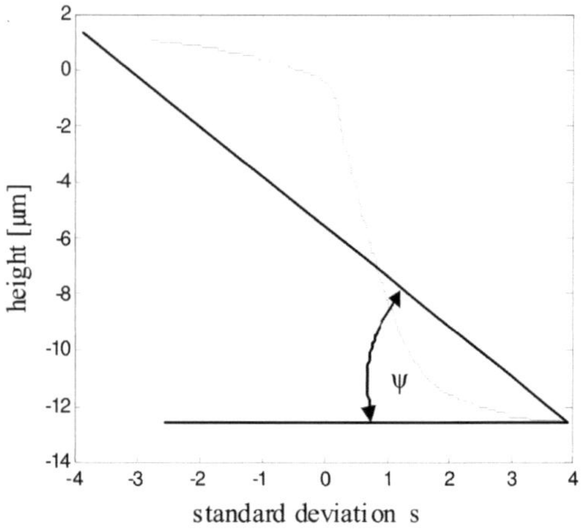

Figure 1. Material probability plot with straight line passing by the first and the finishing point and ψ angle

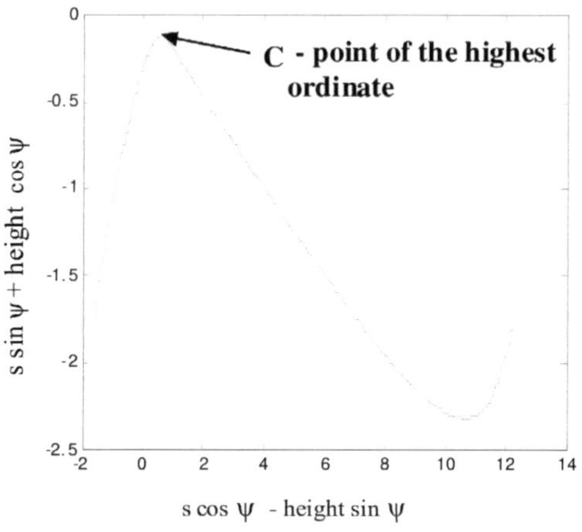

Figure 2. Material probability plot rotated by ψ angle

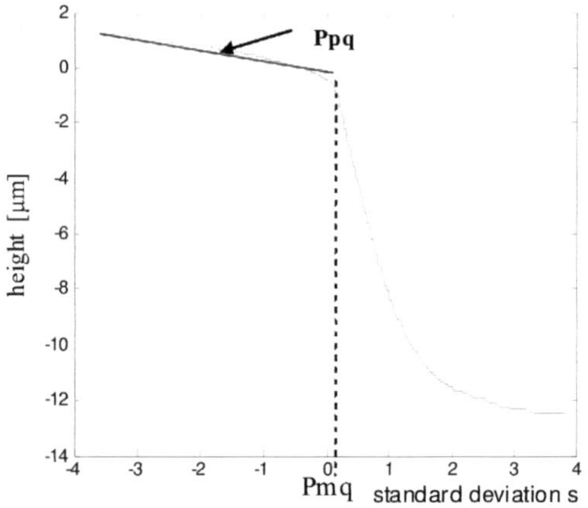

Figure 3. Material probability plot with regression line passing by random region and transition point between random and deterministic regions

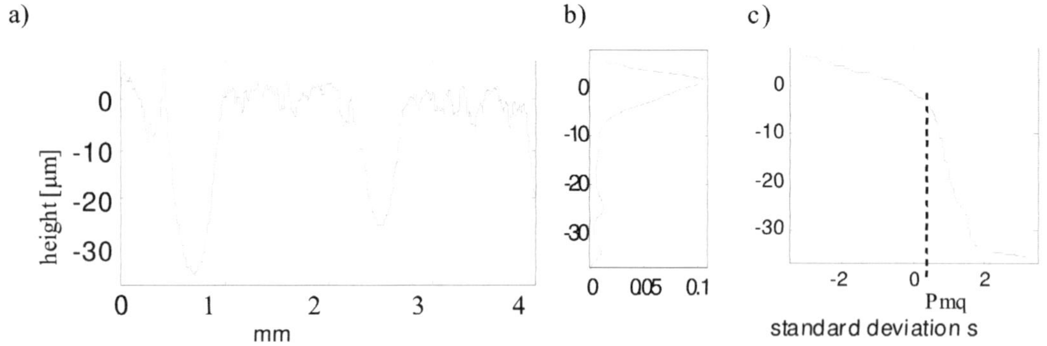

Figure 4. Profile of shaft surface with oil pockets (a), profile ordinate distribution (b), material probability plot (c): Ppq = 2.91 μm, Pmq = 66.85%

Oil pockets density (pit-area ratio) is tribologically important parameter characterizing textured surfaces. It is calculated as ratio of surface area occupied by dimples to the total area. However calculation of pit-area ratio is time consuming. It is necessary to measure area occupied by each dimples (when holes are of spherical shape, determination of mean diameter is sufficient). When the total area is larger than measured area one should determine average area occupied by each dimple, multiply it by the number of holes and divide it by the total area. It is possible to estimate pit-area ratio on the basis of shape of material ratio curve (see Figure 5).

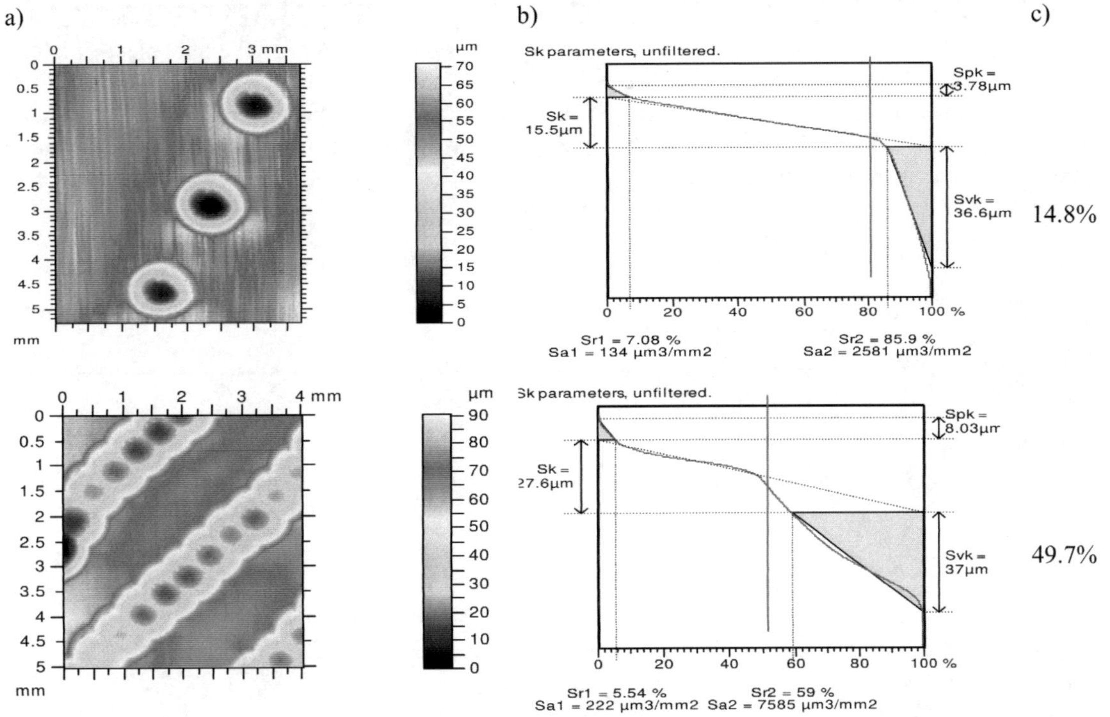

Figure 5. Contour plots of textured surfaces (a), material ratio curves (b), pit-area ratios S_p (c)

This method is quicker than presented above. Because valley part is of deterministic character, therefore it presents straight line on usual (not probabilistic) material ratio curve. Therefore oil pockets density is obtained basing on determination of the point of this straight line finishing. Pit-area ratio S_p is equal 100% - tp, where tp is material ratio (abscissa) of this point. The real area covered by dimples is equal to S_p multiplied by measured area. It is also possible to estimate average area occupied by each dimple. Differently to developed method of the Pqp (Spq) parameter determination, it can be used for surfaces of various character of plateau part (deterministic or random). However there are some problems with estimation of dimples density higher than 70% using the proposed method. Two presented above methods of pit-area ratio S_p estimation were compared for some surfaces with oil pockets created by burnishing technique. The mean difference was 0.44% (the highest was 0.8%). Figure 5 presents examples of oil pockets density determination on the basis of material ratio curve.

3. Selection of reference plane

The initial, but very important operation during analysis of areal surface topography is selection of the reference plane. For improper its choice serious errors in parameter calculation can arise. Only a few technical publications were found in this subject ([11] is the example). The present authors analyzed the effect of degree of a polynomial on areal surface parameters on cylindrical textured elements. The polynomials of degrees from the range 1-12 were analysed. Parameters from standard ISO 25178-2 and oil pockets sizes were studied.

Increase of degree of a polynomial caused decrease of average height parameters Sa and Sq (sometimes the values of these parameters increased when polynomial degree was higher than 10). Peak density Spd and auto-correlation length Sal also decreased when polynomial degree increased. No tendency of texture parameter Str changes was found. However parameters describing maximum surface height St, S10z as well as the core roughness depth Sk, the maximum surface peak height Sp and the lowest valley of the surface Sv initially decreased and increased when the degree of a polynomial was higher than 2 or 3 (depending on the surface) during increase of the degree of a polynomial. Similar tendency of the Sk parameter changes was observed during the analysis of plateau honed surfaces [12]. When the polynomial degree was higher than 2 or 3, areal surface parameters can be improper calculated, and, more importantly the dimples depth could be underestimated, and, as a consequence, the tribologic properties could be falsely assessed (see Figure 6). Therefore the 2nd and 3rd degree of polynomial was recommended.

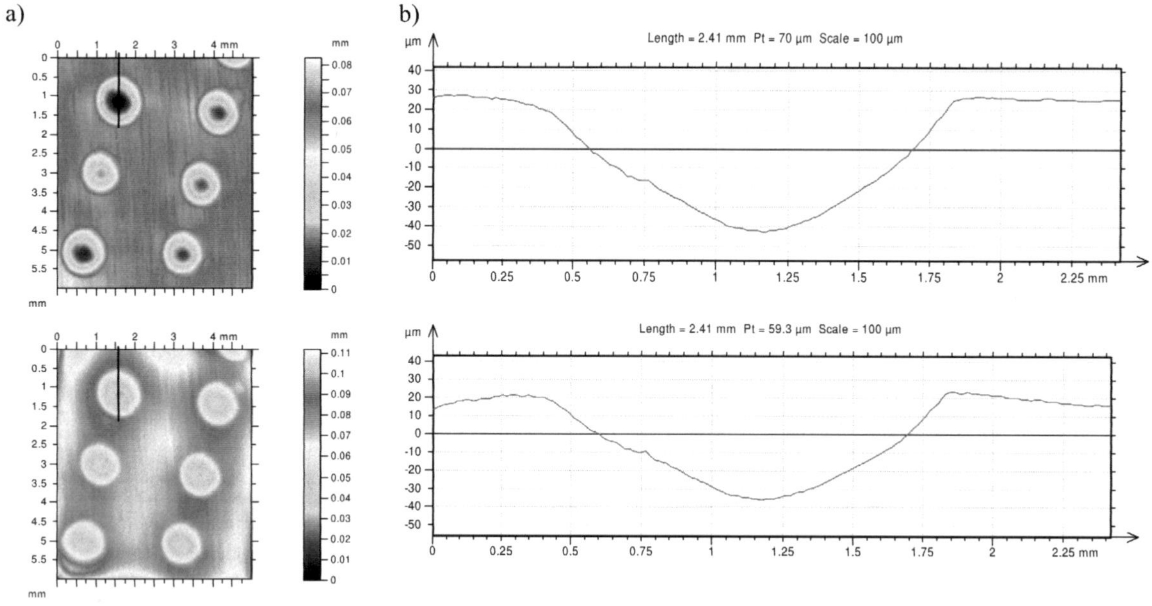

Figure 6. Contour plot of textured surface (a), the shape of oil pocket (b), after application of polynomial of 3rd degree (upper graph) and polynomial of 7-th degree (lower graph) as reference planes

This analysis was done for cylindrical elements of the same diameter. However for textured surfaces the following procedure is proposed: computing parameters for increase of the polynomial degree and selection of this degree, from which the parameters St, S10z, Sp, Sv and Sk started to increase. The information that oil pockets depth if highly correlated (proportional) to St, S10z and Sv parameter values [13] is also important.

Change of the reference element did not affect value of the parameter Std, changes in Sdq, Spd and Sdr parameters were negligible. Figure 6 shows example of false estimation of oil pocket depth as a result of improper selection of the degree of a polynomial (7th).

4. Concluding remarks

Material ratio curve can be applied for determination of tribologically important parameters of two-process textured surface topography.

Material probability curve is useful in determination of standard deviation of plateau height Pqp and abscissa of transition point between deterministic and random part Pmq of surface profiles with oil pockets. The originally developed method for these parameter obtaining is basing on the rotation of material probability plot. Areal extension of the Ppq and Ppq parameters (Spq or Smq) can be also determined in this way. After the analysis of a lot of surfaces it was found that proper parameter values can be obtained using this method.

One can estimate pit-area ratio on the basis of shape of the usual plot of material ratio curve. The oil pockets density can be obtained by determination of the point of finishing the straight line describing deep valley part of deterministic character. This method assures very similar values of pit-are ratio and is much quicker than that commonly used.

After improper selection of textured surface reference plane serious errors in parameter calculation can arise. More importantly, dimple sizes can be assessed falsely. The procedure of increase in the degree of polynomial is proposed, according to it this degree should be chosen, from which the parameters St, S10z, Sp, Sv and Sk started to grow up. For the analyzed surfaces from cylindrical elements of the same diameter the 2nd and 3rd degree of polynomial was recommended. When the polynomial degree was higher the dimples depth could be underestimated. Generally increase of degree of a polynomial caused decrease of the parameters: Sa, Sq, Sal and Spd, however the changes of Std, Sdq, Spd and Sdr were negligible.

5. References

[1] Nilsson B, Rosen B-G, Thomas T R, Wiklund D, Xiao L 2004 Oil pockets and surface topography: mechanism of friction reduction *XI International Colloquium on Surfaces* (Chemnitz: Germany)

[2] Etsion I 2005 State of the art in laser surface texturing *ASME Journal of Tribology* **125** 248-253

[3] Koszela W, Pawlus P, Galda L 2007 The effect of oil pockets size and distribution on wear in lubricated sliding *Wear* **263** 1585-1592

[4] Brinkmann S, Bodschwinna H 2003 Characterisation of automotive bore performance using 3D surface metrology *Advanced Techniques for Assessment Surface Topography* (Kogan Page Science, London and Sterling) pp 307-347

[5] Duffet G, Sallamand P, Vannes A B 2003 Improvement in friction by cw Nd: YAG laser surface treatment on cast iron cylinder bore *Applied Surface Science* **205** 289-292

[6] Klink U 1997 Laserhonen fur Zylinderlaufbahnen *MTZ* **58/9** 554-556

[7] Ryk G, Etsion I 2006 Testing piston rings with partial laser surface texturing for friction reduction *Wear* **261(7-8)** 792-796

[8] Etsion I, Sher E 2009 Improving fuel efficiency with laser textured piston rings *Tribology International* **42** 542-547.

[9] Blunt L, Jiang X, Scott P 2008 Structured surfaces: a new paradigm in surface metrology *XII International Colloquium on Surfaces* (Chemnitz: Germany) pp 403-410

[10] Blunt L, Xiao S 2009 The use of surface segmentation methods to characterize laser zone surface texture og hard disc drives. *XII International Conference on Metrology and Properties of Engineering Surfaces* (Rzeszow: Poland) pp 117-124

[11] Stout K J, Sullivan P J, Dong W P, Mainsah E, Luo N, Mathia T, Zahouani H 1993 The development of methods for the characterisation of roughness in three dimensions – Commision of the European Communities

[12] Swieca I 2010 The effect of reference element on values of areal surface topography parameters, MSc Thesis, Rzeszow University of Techology (in Polish)

[13] Reizer R, Galda L, Dzierwa A, Pawlus P, Simulation of textured surface topography during a low wear process *Tribology International* (in press)

Journal of Physics: Conference Series **311** (2011) 012024 doi:10.1088/1742-6596/311/1/012024

Identification of the multiscale diamond turning signature of optical lens surfaces

S. Mezghani[1], M. El Mansori[1], H. Zahouani[2], G. Martin[3]

[1]Arts & Metiers ParisTech, LMPF, rue St Dominique – BP 508, 51006 Châlons-en-Champagne – France ;
[2]Ecole Centrale de Lyon, LTDS UMR CNRS 5513, 36 avenue Guy de Collongue, 69131 Ecully Cedex, France ;
[3]Essilor international, 81 boulevard Jean-Baptiste Oudry 94000 Créteil, France ;

E-mail: sabeur.mezghani@ensam.eu

Abstract. Functional surfaces of optical lens are commonly achieved by a multi-stage diamond turning. This high precision process acts in a wide range of wavelength and allows producing 3-dimensional free form optical surfaces with excellent surface finish that meets application requirements. However, the relationships between process variables and surface characteristics are not yet predictable. In this paper, the concept of the multiscale process signature (MPS) is applied to track the effect of diamond turning process variables (cutting velocity, feed rate, cutting depth, tool roughness,...) on the surface topography from micro-roughness to waviness. The MPS is developed based on continuous wavelet transform and it depicts the essential changes of the surface state produced on the original surface after diamond turning. Using this concept, the effects of different working variables are isolated and theirs active wavelength bands were identified.

1. Introduction

The functional 'Mirror' surface of ophthalmic plastics lenses is actually produced by multi-processes, turning and polishing [1]. The turning stage plays an important role in the production process and it has two main functions [2]. First, it allows achieving the expected optical surface in terms of geometry to obtain the right light rays deviation for medical eyesight correction; Secondly, it has the important task to make the surface compatible with the post-polishing scale to remove tool marks and to get the desired level of transparency. Hence, the monitoring of the turning process is essential because this process is influenced by lot of working variables such as tool wear, feed rate, and depth of cut [3, 4]. An imprint of all the static as well as dynamic forces, stresses and strains during the cutting process resulting from the operating conditions are left in the generated surface profile in wide range of wavelength. Then, an accurate analysis of the surface features dynamic is necessary to extract the process signature [5].

In this work, the concept of multiscale process signature (MPS) [6, 7] is applied as a new experimental investigative approach for monitoring the dynamics of surface generation during single-point diamond turning of ophthalmic optical lens. As we will see in subsequent sections, the multiscale approach acts as a powerful indicator to extract the turning process signature. It also helps monitoring of the diamond tool wear and the scale of turning process impact in relation to the process working variables.

In particular, the effect of the cutting tool acuity is demonstrated in enhanced finishing of optical surface that will conditioning the generation of 'Mirror' surface after polishing.

2. Experimental procedure

In this work, single-point diamond turning experiments were carried out on industrial machine (Satisloh VFT-compact) (figure 1). Table 1 show machining conditions used in production line of ophthalmic lenses. In this process configuration, surface of ductile cutting is generated.

Figure 1. Photograph of the single-point diamond turning machine

Table 1. Single-point diamond turning experiments conditions

Turning stage	Rough	Finish
Lens workpiece material	Polycarbonate	
Rake angle of the tool	0°	0°
Tool nose radius r_ε *(mm)*	1.7	2.2
Spindle speed V_r *(rpm)*	5000	2000
Feed amount f *(µm)*	52	66
Depth of cut a *(mm)*	2	0.2

The workpiece consists of aspherical polycarbonate lenses with diameter equal to 75mm. The turning process is started by the movement of the cutting tool along the z-direction by the z-carriage of the diamond turning machine until a defined depth of cut (a) of the lens array. During the process, the spindle is moved along the x-direction by (f) feed step (Figure 1). Each turning test was repeated five times. Three topographic profiles were automatically recorded with a mechanical stylus on a SURFASCAN (Hommel-Somicronic) profilometer at different location on the lens sample diameter, before and after each turning stage.

3. Concept of the multiscale process signature (MPS)

To derive the quantitative expression for each stage of turning process, a multiscale analysis of the surface texture is considered which involves the decomposition of its topographic profiles into wide range scales. This decomposition uses continuous wavelet transform which can be considered as a mathematical microscope, where the resolutions are the basic functions obtained from a single wavelet or mother wavelet $\psi(x)$ by dilation (or compression) and translation [6, 7]. The "Morlet 1D" wavelet given by the following expression is used:

$$\psi(x) = \pi^{-1/4} e^{iw_0 x} e^{-x^2/2} \tag{1}$$

The result of the decomposition makes it possible to identify the various scales of the topographic signal after a 1D inverse wavelet transformation [6]. The methodology consists of the quantification of the arithmetic mean value for each scale. The objective is to determine a quantitative expression of the spectrum of arithmetic mean value from the scales of waviness to roughness, Ma [6, 7]:

$$Ma(a) = \frac{1}{N}\sum_{x=1}^{N}\left|W_a^*(x)\right| \tag{2}$$

A transfer function of the process that characterizes the roughness attenuation at each scale is introduced. It's obtained by determining the ratio of SMa(a) spectrum between the surface irregularities before and after finishing (Equation 3).

$$MPS(a) = \frac{Ma_F(a) - Ma_I(a)}{Ma_I(a)}.100 \tag{3}$$

With $Ma_I(a)$ and are $Ma_F(a)$ respectively the roughness spectrum of the surface before and after the considered turning stage.

This computed transfer function constitutes the multiscale process signature. It identifies the effect of machining at any scale, and permits quantification of the quality of the surface finishing at defined wavelength band [6, 7].

4. Identification of the multiscale single-point diamond turning process signature

Figure 2 shows the MPS spectrums of the rough and finish turning stage respectively. It show very clearly the occurrence of a maximum (minimum of surface irregularities attenuation) located at wavelength value equal to the feed amount (f) (figure 3).

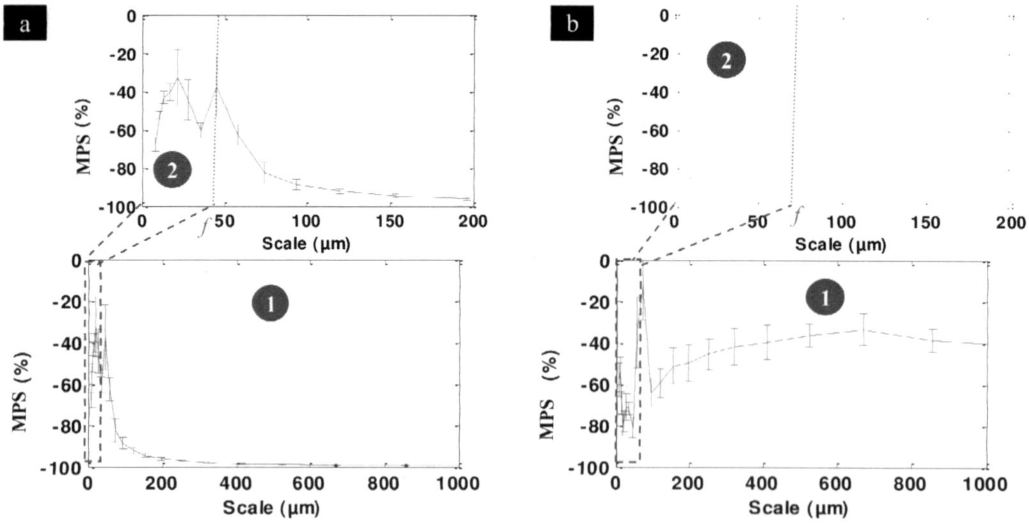

Figure 2. MPS of the single-point diamond turning process: (a) rough turning and (b) finish turning

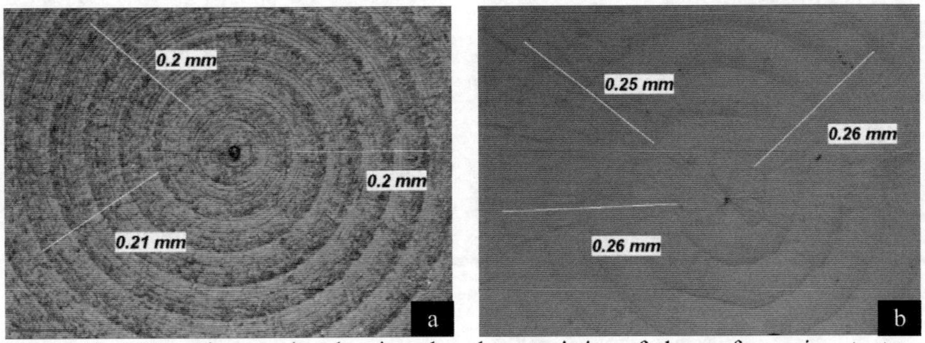

Figure 3. SEM micrographs showing the characteristics of the surface micro-texture generated by: (a) rough turning and (b) finish turning stage.

This maximum of the MPS represents the turning straight pattern and separate the wavelength range into two domains:

① for $scale \geq f$, MPS spectrums of rough and finish turning stage present respectively different attenuation rate of surface irregularities. This attenuation rate reaches the 95% at the first turning stage for all scale greater than 200µm. This suggests that the cutting action in rough turning stage occurs in a wide scale range due to a greater contact between the tool and the machined surface. The cutting conditions for each turning stages calculated by using the model developed by Dong et al. in [8] (see Table 2) confirm that the cutting section area Ac in the rough turning step is almost eight times higher than in the finishing stage.

② for $scale \leq f$, as we can see from the MPS spectrum (figure 2), the finish turning stage leads to a higher reduction of surface micro-irregularities. The MPS in this scale domain seems to be independent from cutting variables as in metallic turning. In fact, in this scale domain, if we consider as an example, the MPS of the rough turning stage shows a maximum located at almost 18 µm. This implies the presence of elementary patterns with 18 µm of size on the lens surface. The topography of cutting tool edge shows an autocorrelation length of 17.5µm (figure 4-b). Hence, the micro-feature on the topography of the rough turning tool and on the generated surface has almost the same size. Then, it represents a fingerprint of the turning tool roughness.

Table 2. Cutting conditions computed by the model described in [8]

Turning stage	Rough	Finish
Cutting section area (Ac) (mm^2)	0.1040	0.0132
Cutting width (aw) (mm)	3.6736	0.9695
Cutting thickness (ac) (mm)	0.0283	0.0136

Hence, the functional impact of the turning process can be entirely described by four parameters that can be used as inputs for the optimization of the post-polishing stage. These parameters are:

- The maximum amplitude of the MPS spectrum;
- The scale of the maximum amplitude of the MPS spectrum (f); it define the scale of the turning marks to be removed by polishing;
- MPS rate average for $scale \leq f$;
- MPS rate average for $scale \geq f$;

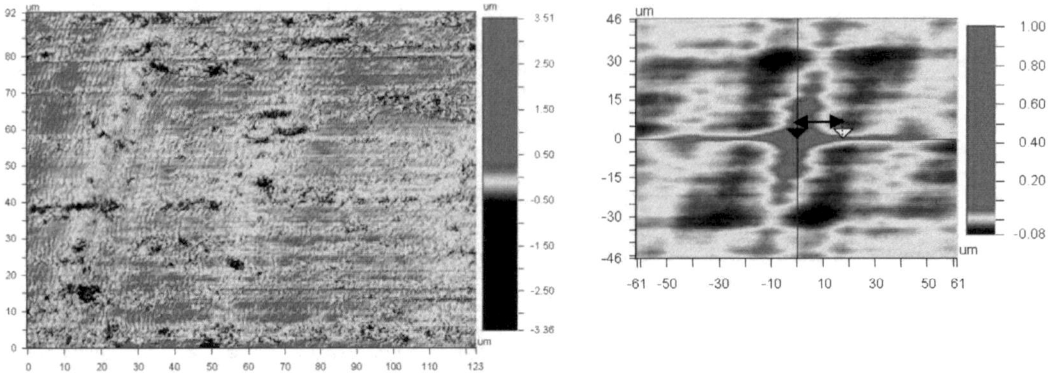

Figure 4. Surface topography of the turning tool used in the rough turning stage (a) and its auto-covariance function (b) showing a correlation length of 17.5 µm in the machining direction.

5. Conclusion

A new investigative approach to monitor the dynamic of surface generation can be used to increase the level of mastery of the single-point diamond turning process. It is capable of throwing light on both macroscopic and microscopic performances of each turning stage.

The conclusions reached from the analysis of the machined surfaces within the industrial operating conditions were as follow:

- Small wavelength components of the roughness, with wavelength less than the feed working variable are independent from cutting parameters. It depends only on the micro-geometry of the tool and hence the mean average of the MPS spectrum in this scale domain can be used as a parameter to monitor the tool life.
- The reduction of surface irregularities of large wavelength components greater than the feed working variable is considerably dependent on the ratio between the depth of cut and tool nose radius (a/r_w).

Overall, this approach can help in the production of high effective cutting tool. This approach is also necessary for the identification of the optimal MPS of turning which can improve the efficiency of the post-polishing stage in relation to the working process variables.

References

[1] J-D Kim, I-H Choi, "Ultrasonic vibration turning of optical plastics", Proc. of SPIE Vol. 2576, 1995, pp. 414-425.

[2] W. Jiang, "Diamond turning microstructure optical components", 4th international symposium on Advanced Optical Manufacturing and Testing Technologies: Advanced Optical Manufacturing Technologies, edited by Li Yang, John M. Schoen, Yoshiharu Namba, Shengyi Li, Proc. of SPIE Vol.7284, 2009, pp. 1-5.

[3] A.G. Mamalis, S.N. Lavrynenko, "On the precision single-point diamond machining of polymeric materials, Journal of Materials Processing Technology 181 (2007) 203-205.

[4] Y.J. Noh, Y. Arai, W. Gao, « Improvement of a fast tool control unit for cutting force measurement in diamond turning of micro-lens array », Int. J. Surface Science and Engineering, Vol. 3 (3), 2009, pp. 227-241.

[5] S.M. Pandit, S. Revach, "A data dependent systems approach to dynamics of surface generation in turning", journal of engineering for industry 103 (1981), pp.437-445.

[6] H. Zahouani, S. Mezghani, R. Vargiolu, M. Dursapt, "Multi-scale study of abrasion signature by

2D wavelet decomposition", Wear, Vol. 264 (5-6), 2008, pp. 480-485.

[7] M. El Mansori, S. Mezghani, L. Sabri, H. Zahouani, "On the concept of process signature in the analysis of multi-stage surface formation", Surface Engineering, 26 (3), 2010, pp. 216-223.

[8] G. Dong, C. Han, S. Dong, T. Sun, "Influence of cutting section and corner radius to cutting force in diamond turning", 4th international symposium on Advanced Optical Manufacturing and Testing Technologies: Advanced Optical Manufacturing Technologies, edited by Li Yang, John M. Schoen, Yoshiharu Namba, Shengyi Li, Proc. Of SPIE Vol.7282, 2009, pp. 1-5.

Journal of Physics: Conference Series **311** (2011) 012025 doi:10.1088/1742-6596/311/1/012025

On the optimal choice of wavelet function for multiscale honed surface characterization

S. Mezghani[1], L. Sabri[2], M. El Mansori[1], H. Zahouani[3]

[1]Arts & Metiers ParisTech, LMPF, rue St Dominique – BP 508, 51006 Châlons-en-Champagne – France ;
[2] RENAULT S.A.S., Direction de la Mécanique/Direction de l'Ingénierie Process, Rueil Malmaison (Paris), France
[3] Ecole Centrale de Lyon, LTDS UMR CNRS 5513, 36 avenue Guy de Collongue, 69131 Ecully Cedex, France ;

E-mail: sabeur.mezghani@ensam.eu

Abstract. Multiscale surface topography characterization is mostly suited than standard approaches because it is more adapted to the multi-stage process generation. Wavelet transform represents a power tool to perform the multiscale decomposition of the surface topography in a wide range of wavelength. However, characterization results depend closely on the topography data acquisition instrument (resolution, height accuracy, sensitivity…) and also on the wavelet analysis method (discrete or continuous transform). In particular, the choice of wavelet function can have significant effect on the analysis results. In this paper, we present experimental work on a number of popular wavelets functions with the aim of finding wavelets that exhibit optimal description of honed surface features when continuous wavelet transform is used. We demonstrate that the regularity property of wavelet function has a significant influence on the characterization performances. This comparative study shows also that the Morlet wavelet is the more adapted wavelet basis function for multiscale characterization of honed surfaces using continuous wavelet transform.

1. Introduction

Faced with the development of several varieties of wavelet functions, several studies in various applications have tried to identify the effect of the wavelet function on its performances. Indeed, since the wavelet transform involves correlating the signal being analyzed and a prototype wavelet function, the choice of the wavelet function may influence the performance of wavelet decomposition. Actually, there is not yet a general answer for all wavelet application. In fact, for example, in textures classification applications, several comparative studies conclude that the choice of the mother wavelet function has little effect on the results [1, 2]. While, other studies show that the analysis of cardiac signals is optimal when using "Meyer" wavelet [3]. In data compression applications, the choice of wavelet is not very critical as long as it is reasonably smooth [4]. So, the choice of the "best" mother wavelet is obviously application dependent. This suggests that wavelet function must be adapted to the characteristics of relevant patterns that differ from one application to another. According to Wickerhauser [5, 6], two desirable properties for adapted wavelet bases are necessary. First, the spatial localization of each feature patterns at each scale particularly in high scale (low frequency).

Journal of Physics: Conference Series **311** (2011) 012025 doi:10.1088/1742-6596/311/1/012025

The second property is the frequency localization to guaranty the separation of multiple features into different scales.
In this paper, various wavelet functions within a surface topography characterization approach based on 1D Continuous Wavelet Transform (CWT) are tested on honed topographic profiles. The effect of the mother wavelet choice is at first pointed out. Then, the capacity of each wavelet to provide the best frequency-space localization is highlighted. Their sensitivity to the roughness amplitude is also investigated. Optimal wavelet function for honed surface characterization using continuous wavelets transform is then identified.

2. Continuous wavelets-based characterization approach

We give here a brief description of wavelet based characterization approach. This decomposition used 1D continuous wavelet transform. The continuous wavelet transform can be interpreted as a multi-channel filter system. The surface topography components pass through a filter bank which is a set of the contracting wavelets. The number of wavelets corresponds to the number of iterations.

One defines the 1D continuous wavelet transform of a topographic signal f(x) by [7, 8]:

$$W_{b,a}(x) = \frac{1}{\sqrt{a}} \int_{-\infty}^{+\infty} f \; x \, \psi(\frac{x-b}{a})dx \qquad (1)$$

Where “a” is the contraction coefficients according to the x-direction, b the translation coefficients according to the x-direction.

The result of the decomposition makes it possible to identify the component of the surface at each scale after a 1D inverse wavelet transformation. The superposition of all scales of the decomposition creates a plane in (x, scale) basis. The characterization methodology consists of quantifying the arithmetic mean values at each scale of decomposition. A spectrum of roughness absolute magnitude “Ma” is obtained.

3. Effect of the “Mother” wavelet function on honed surface characterization

To determine the optimal wavelet function for the multiscale characterization of honed surfaces, several wavelet basis functions are considered namely, Morlet, Meyer and Gaussian derivative at different order. Figure 1 shows of wavelets families’ functions.

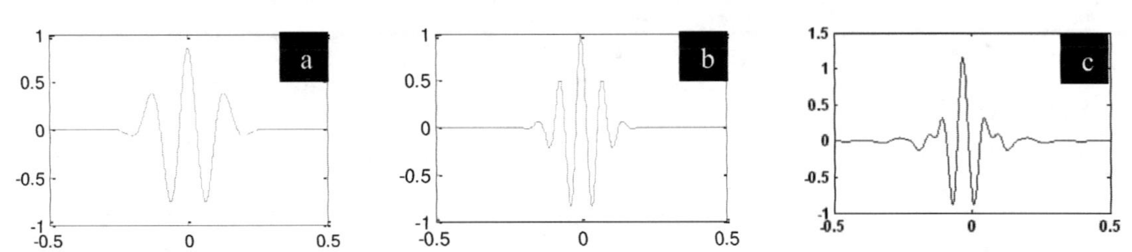

Figure 1. Examples of wavelets families’ functions expressed in the spatial domain (a) 8th derivative Gaussian (DOG8), (b) Morlet et (c) Meyer.

Then, topographic profiles of surfaces after finish honing are considered. They are sampled in four thousand points with step scale of 4µm. Each signal is decomposed by CWT in thirty two scales.
Figure 2 shows the Ma spectrum obtained by the characterization approach described above and using different wavelet functions. Although all the applied wavelet functions lead to a low reconstruction error, the multiscale Ma spectrum vary from a wavelet basis function to another. The choice of the wavelet "Mother" has then a major influence on the multiscale characterization of surface features.

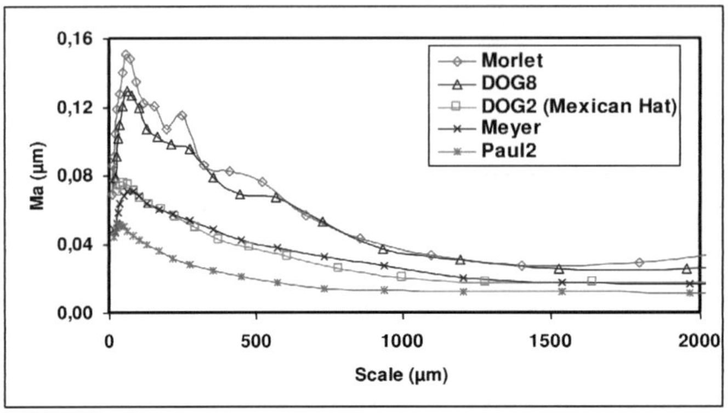

Figure 2. Ma Spectrum of finish honed surfaces obtained using different Mother wavelet functions.

4. Identification of the optimal wavelet function choice

To find the Mother wavelet that leads to a complete and precise description of the honed surface topography, we must determine the wavelets family allowing precise detection of all elementary patterns of honed surfaces. This simply boils down to analyzing the spatial and frequency localization of the topographical signal patterns. To qualify the detection and localization quality of a given wavelet function, we identify firstly the basic and repeatable pattern shape of a honed surface (figure 3-b). The identified characteristic pattern represents the indentation trace left by one abrasive grit on the surface of the cylinder. It is common to all abrasive finishing processes. This suggests that the results of the wavelet choice study can be generalized to other abrasive processes (belt finishing, polishing ...).

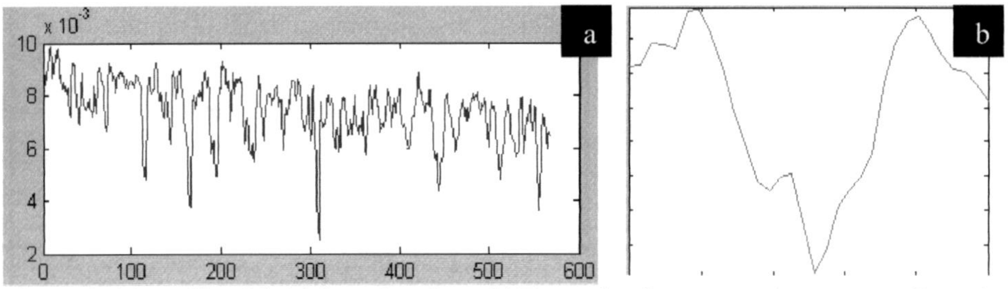

Figure 3. (a) Profile of finish honed surface, (b) Example of representative pattern of honed surface.

A signal $t(x)$ is generated for testing the performance of the different wavelet families (Fig. 3-b). This signal is composed by two elementary patterns in two different positions and at two different scales and it is defined by the following expression:

$$t(x) = f((x-b_1)/a_1) + \sqrt{2} f((x-b_2)/a_2) \qquad (2)$$

Then, two local maxima would be detected on the space-scale diagram of the wavelet transform:

- The first in the position $b_1 = 20$ and the scale $a_1 = 8$;
- The second in the position $b_2 = 40$ and the scale $a_2 = 4$;

Journal of Physics: Conference Series **311** (2011) 012025 doi:10.1088/1742-6596/311/1/012025

Besides, honed surfaces exhibit different levels of roughness at each step of honing process. To qualify the sensitivity to the detection of the honed surface patterns in the presence of plateaus micro-roughness, an additional random noise is added to the signal $t(x)$ (Figure 4-c).

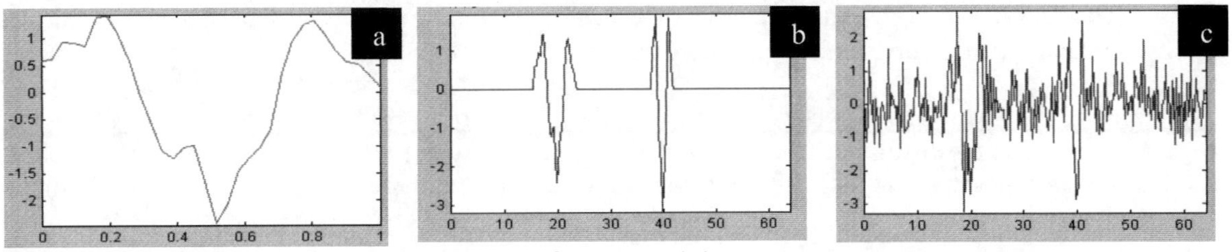

Figure 4. (a) Elementary motif of honed surface $f(x)$ (b) $t(x)$ signal constituted by two motifs at two scales in two distinguishes positions.

4.1. Spatial and frequency localization performance

As it was already mentioned, by means of the wavelet choice we can access different time and frequency resolution in the time-frequency plane. This is demonstrated here for the topographic signal using various wavelet functions. Figure 5 shows the coefficients of the time-frequency representation of the wavelet transform of the signal $t(x)$ using different wavelet functions. It summarizes the space-frequency localization performance.

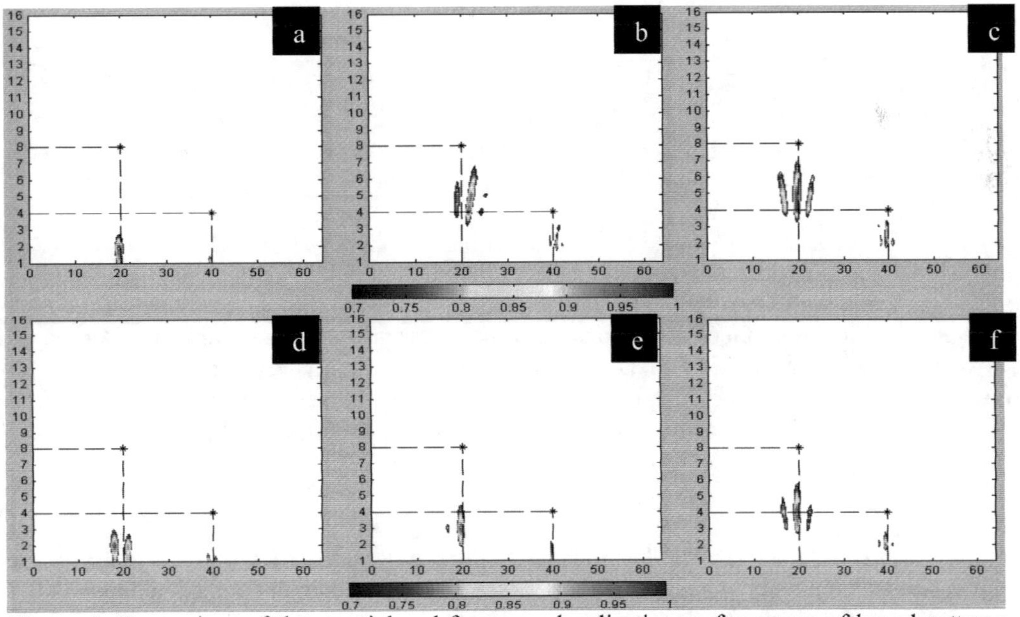

Figure 5. Comparison of the spatial and frequency localization performances of honed patterns using different wavelets functions : (a) Mexican Hat, (b) Meyer, (c) Morlet, (d) 1st derivative of Gaussian function, (e) 4th derivative of Gaussian function, (f) 8th derivative of Gaussian function.

By comparing these different wavelets coefficients planes, we can state that:

- The "Mexican Hat" and "Meyer" wavelet are not suited because they are not adapted to detect honed surfaces patterns (Fig. 5-a). For example, the pattern located in b2=40 should be detected at scale 4 does not appear in the wavelet coefficients spectrum.
- The "Morlet" wavelet and 8th order of Gaussian wavelet function present the more accurate robustness for honed surface patterns detection.

Furthermore, the "Gaussian" wavelets studied with three derivative orders (two, four, and eight). With higher filter order, the regularity of wavelets increases and in turn increases the compaction energy and the number of vanishing moments. Results shows that better frequency localization are obtained by applying more regular wavelet (figure 5). Thus, the order of a wavelet is important in achieving a representative characterization.

4.2. Sensitivity to the roughness amplitude

Figure 6 shows, by comparison of maxima amplitude of wavelet coefficients, that honed patterns detection using Morlet wavelet is less sensitive to the micro-roughness than by using the 8th Gaussian derivative wavelet function.

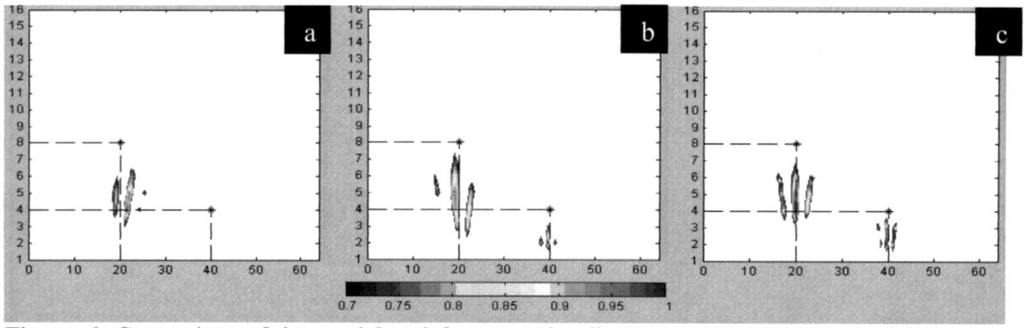

Figure 6. Comparison of the spatial and frequency localization performances of honed patterns using different wavelets functions: (a) Meyer, (b) 8th derivative of Gaussian function, (c) Morlet.

5. Conclusion

The presented work highlights the importance of the choice of the wavelet basis function in the characterization of honed surfaces. It demonstrates that the regularity is a fundamental property in the analysis of topographic profiles from honed surfaces. Hence, more regular (higher order) wavelet functions yield to precise detection of honed surface patterns. Based on this observation, Morlet wavelet appears as the most appropriate wavelet function for analyzing such signals.

References

[1] S. M. Debbal and F. Bereksi-Reguig, Choix de l'ondelette analysante et classification des signaux phonocardiogram m es en fonction des souffles surajoutés, Afrique SCIENCE 01(1) (2005) 1 – 13 (ISSN 1813-548X)

[2] Misiti, M.; Y. Misiti, G. Oppenheim, J.-M. Poggi (2003), "Les ondelettes et leurs applications," Hermes.

[3] A.O. Rodriguez, P. Mansfield, wavelet enhancement method to visualize heart blood flow, Proceedings of SPIE - The International Society for Optical Engineering, Vol. 3979, pp. I/- ,2000.

[4] L. Ke, Q. Du, Z. Su, The correlation between the wavelet base properties and image

compression, Proceedings - CIS Workshops 2007, 2007 International Conference on Computational Intelligence and Security Workshops, N° 4425526, pp. 429-432, 2007

[5] V. Wickerhauser, Adapted Wavelet Analysis from Theory to Software, AK Peters, Boston, 1994.

[6] N. Ahuja, S. Lertrattanapanich, K. Bose, Properties determining choice of mother wavelet, IEE Proc., Vis. Image Process., Vol. 152 (5), pp. 659-664, 2009.

[7] H. Zahouani, S. Mezghani, R. Vargiolu, M. Dursapt, "Multi-scale study of abrasion signature by 2D wavelet decomposition", Wear, Vol. 264 (5-6), 2008, pp. 480-485.

[8] M. El Mansori, S. Mezghani, L. Sabri, H. Zahouani, On the concept of process signature in the analysis of multi-stage surface formation, Surface Engineering, 26 (3), pp. 216-223, 2010.

Journal of Physics: Conference Series **311** (2011) 012026 doi:10.1088/1742-6596/311/1/012026

Multiscale effect of paint pulverization orientation on appearance after painting

S. Mezghani[1], H. Zahouani[2], J. Piezanowski[3]

[1]Arts & Metiers ParisTech, LMPF, rue St Dominique – BP 508, 51006 Châlons-en-Champagne – France ;
[2]Ecole Centrale de Lyon, LTDS UMR CNRS 5513, 36 avenue Guy de Collongue, 69131 Ecully Cedex, France ;
[3]Arcelor Ledeep Florange, 17 Avenue des Tilleuls, 57191 Florange, France;

E-mail: sabeur.mezghani@ensam.eu

Abstract. The perceived quality of a vehicle is strongly affected by paint appearance that shares major part of the outer car body panels. The painting process modifies the surface topography in a wide range of roughness and waviness scales, and consequently modifies the functionality of the surface in terms of appearance. Since painting process is a multistage process leading to stratified surfaces, a multiscale surface topography characterization approach is suited. In this paper, 2D multiscale signature of the painting process was introduced and applied to track the effect of the painting process working variable on painted surface topography in a wide range of wavelength. To this aim, experimental painting tests were performed using three painting orientation modes (horizontal, oblique and vertical) on random and deterministic metal sheet surface textures. Results show that the painting orientation mode affect only the wavelength band greater than 500 μm and optimal painting orientation depends strongly on the texture of the initial sheet surface.

1. Introduction

The exterior paint finish of a car plays an important part in how the customers perceive its quality. Unfortunately, the large number of different process parameters (paint thickness, substrate texture and orientation during paint application...) makes it complicated for the plant operator to maintain a high quality of production. Moreover, quality inspection and monitoring are carried out visually these days, by means of measurements by trained personal after the painting process [1]. This method is inadequate for various reasons: (1) undefined evaluation conditions; (2) judgment is dependent on the mood of the observer; and (3) different observers have different visual acuity [2, 3]. Therefore, improved quality inspections and objective quality criteria are necessary to meet rigorous quality requirements.

In this paper, the concept of multiscale process signature (MPS) is applied as a new experimental investigative approach for monitoring the dynamics of surface features modification after paint coating with different substrate orientations [3, 4]. To this aim, experimental painting tests were performed using three painting orientation modes (horizontal, oblique and vertical) on random and deterministic metal sheet surface textures. Rays-tracing simulations of light scattering from painted surfaces were performed to track the effect of substrate orientations during painting on surface appearance. As we will see in subsequent sections, MPS spectrum presents a good indicator of the light scattering behavior of the painted surface and hence of its appearance.

2. Experimental procedure

The paint finishing of automotive board is a multistage process [2, 3]. It consists of the deposition of three successive paint layers. The first coated layer called "Primer" is the thickest; its thickness varies between 37 μm and 43 μm. The second coat is the main layer of the paint, we call it "basecoat" and it has a thickness varying between 32 μm and 38 μm. Finally, the last top coat "Lacquer" is a finer film, ranging from 14 μm to 16μm thicknesses.

In this study, three orientations of the substrate sheet surface during painting were considered: horizontal, oblique (45°), and vertical. In addition, two main different textures types of a steel sheet substrate were considered: stochastic and deterministic (see Figue 1). They were finished by a succession of the three paint coating layers. Fig. 2 shows the topography of the steel substrate of each texture categories at the final stage by three orientations of the substrate during painting.

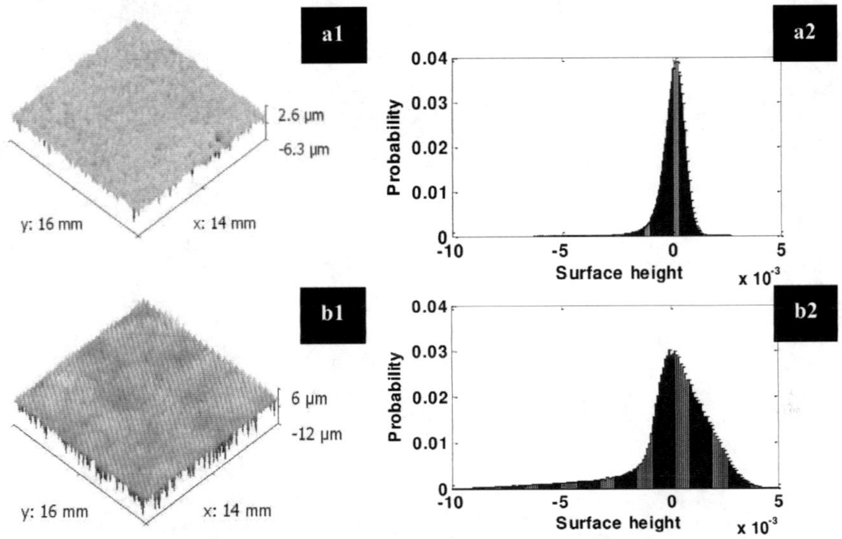

Figure 1. 3D metal sheet surface topography with two texture categories: (a1) random and (b1) deterministic and their height distribution histogram respectively (a2) and (b2).

In each painting process configuration, the measurement of the initial and painted surface topography was carried out by interferometric instrument (Interferometer NT 2000 WYKO). This 3D microscope can be prided on a resolution of 2 μm and total amplitude of measurement bordering 2mm.The resolution in z-direction of interferometer is 10 nm. The surface was sampled in 512×512 points with a 26 μm step scale in the x-direction and 30 μm in the y-direction. In this study, the wavelength bandwidth range from 200 μm to 6 mm is considered.

3. Multiscale quantitative expression for the scale of painting appearance

To derive a quantitative expression for paint finishing, we consider a multiscale analysis of the surface texture before and after painting which involves the decomposition of its topographic profiles into different roughness scales. This decomposition uses continuous wavelet transform which can be considered as a mathematical microscope, where the resolutions are the basic functions obtained from a single wavelet or mother wavelet by $\psi(x)$ dilation (or compression) and translation [3, 4]. The "Morlet 1D" wavelet given by the following expression is used:

$$\psi(x) = \pi^{1/4} e^{i\omega_0 x} e^{-x^2/2} \tag{1}$$

The result of the decomposition makes it possible to identify the various scales of the topographic signal after a 1D inverse wavelet transformation [4]. The methodology consists of the quantification of the arithmetic mean value for each scale.

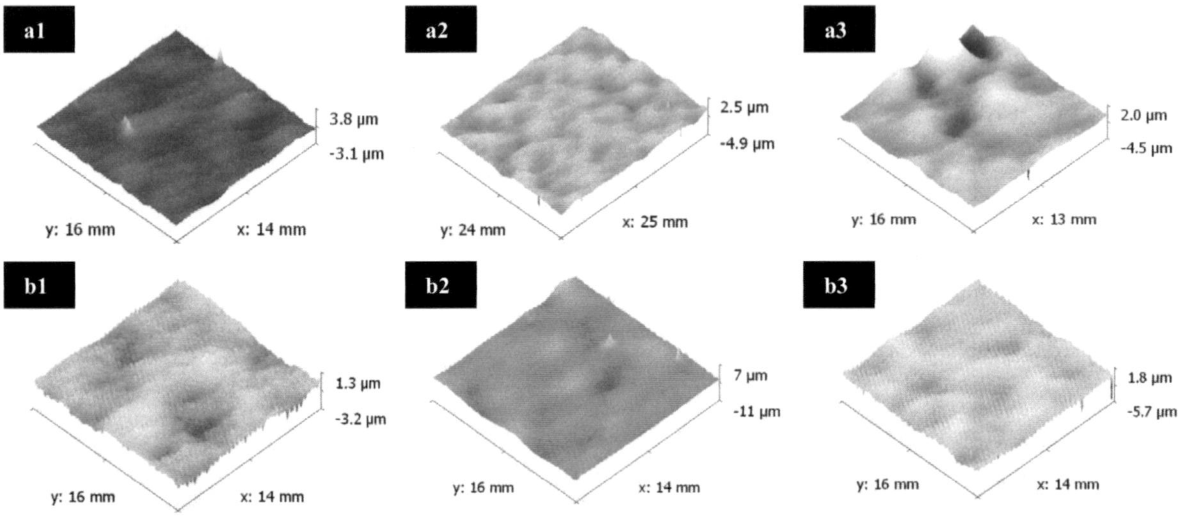

Figure 2. 3D topography respectively of random (a) and deterministic (b) substrate surface texture obtained by three orientation modes during painting process: (x1) horizontal, (x2) oblique 45°, and (x3) vertical.

The objective is to determine a quantitative expression of the spectrum of arithmetic mean value from the scales of waviness to roughness, Ma [3, 4]:

$$Ma(a) = \frac{1}{N}\sum_{x=1}^{N}\left|W_a^*(x)\right| \tag{2}$$

The multiscale characterization methodology was applied in order to quantify the effect of the painting orientation mode on the various scales of the substrate surface. The painting process can be expressed by the action of a weight function, which transforms all the scales of the initial surface (Fig. 1). The painted surface F(x, y) and the initial substrate topography I(x, y) can be related by the multiscale process signature MPS(a) of painting using Eq. (3) [3, 4].

$$MPS(a) = \frac{\left[SMa^F(a)\right] - \left[SMa^I(a)\right]}{\left[SMa^I(a)\right]}.100 \tag{3}$$

where $SMa^I(a)$ and $SMa^F(a)$ are respectively the multiscale roughness spectrum of the surface before and after painting.

The MPS spectrum of the painting process between the final stage (Top coat) and the initial surface topography for three different position modes (horizontal, oblique (45°), and vertical) were computed. Figures 3-a and 3-b show the influence of the sheet substrate orientation during painting on the generated surface topography, respectively, for stochastic and deterministic texture of the initial sheet.

We can remark from these figures that structured surfaces with small size patterns (0.35 mm) permit to discard generation of waviness scales greater than 3mm. Moreover, The MAF spectrum shows that substrate orientation working variable affects the waviness scales. In fact, independently of the initial surface texture, horizontal painting mode leads to more waviness attenuation than by oblique (45°) and vertical mode.

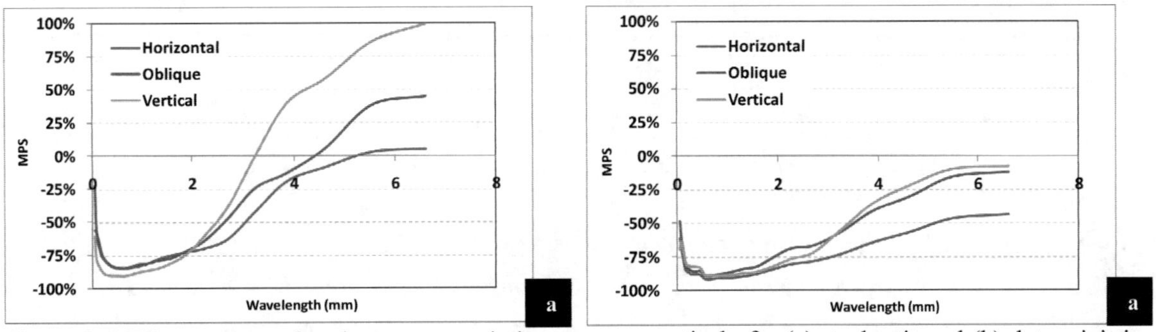

Figure 3. MPS spectrums after the top-coat painting stage respectively for (a) stochastic and (b) deterministic surface texture of steel sheet substrate. Three substrate orientations (Horizontal, oblique and vertical) during the paint process are considered for the two kinds of initial surface texture.

4. Identification of the multiscale single-point diamond turning process signature

In this section, we are interested to investigate the light scattering from surfaces after painting process with different substrate orientation during painting. The objective is demonstrating the pertinence of the MPS as multiscale indicator of surface appearance in relation to the process working variables.

4.1. Ray-tracing simulation tests

The simulation is based upon geometrical optics approximation of the exact numerical integration methods available from electromagnetic wave theory. The approximation is ray-tracing approach, where energy bundles are traced throughout their interactions with the surface until they leave it [5-7]. The geometrical optics approximation (GOA) is valid for $\sigma*\cos(\theta)/\lambda > 0.17$ and for $\sigma/\tau < 2.0$, where (σ) is the root mean square roughness height, (τ) is the correlation length, (θ) is the angle of incidence and (λ) is the light wavelength [5].

In this work, 500 rays were considered in each simulation allowing a tradeoff between execution time and statistical significance of the simulation results. Assuming that the whole surface is illuminated, the beam radius is equal to the surface length. A typical human eye will respond to wavelengths from about 390 nm to 750 nm. A light-adapted eye generally has its maximum sensitivity at around 555 nm. At this wavelength, the refractive index of the basecoat is considered equal to 1.5.

We assume here that the top-coat layer is opaque. Hence, it is where all the light is reflected. The root mean square of height and correlation length in the x and y direction of each painted surface are used as input in the simulation test. These parameters are computed from the acquired surface topography of each surface.

4.2. Effect of painting process orientation on light

Figure 4 show the influence of the sheet substrate orientation during painting on the bidirectional reflectance distribution function (BRDF) for an incidence angle of 20°. The BRDF spectrums represented in figure 4 show that horizontal painting mode leads to a surface topography characterized by the most important light scattering in the specular direction (fig. 4). Then, the best paint appearance is obtained when the sheet substrate is oriented horizontally during the painting process. From the MPS spectrum given in figure 3, we can see that this painting mode (horizontal orientation mode) leads also to the greater waviness attenuation in particular for wavelength greater than 2 mm (fig.3) and independently to the initial substrate texture. Therefore, the better surface appearance results from the greatest waviness reduction.

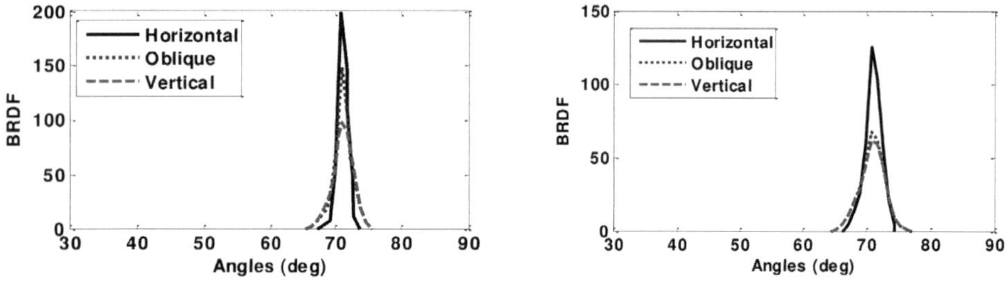

Figure 4. BRDF of the different operating conditions respectively for: (a) stochastic and (b) deterministic substrate surface texture.

5. Conclusion

In this work, the concept of multiscale process signature (MPS) is applied as a new experimental investigative approach for monitoring the effect of substrate orientation during the paint coating process. Rays tracing simulations of light scattering from painted surfaces were performed to track the effect of substrate orientations during painting on surface appearance. Results demonstrate that the appearance quality of painted surface is closely related the waviness amplitude particularly for wavelength greater than 2 mm. Therefore, MPS spectrum ensures the link between the operating parameters of the painting process and the appearance quality of the generated surface.

References

[1] Talbert Roger, Paint technology handbook (ISBN 978-1-57444-703-3), p.172.

[2] H. Zahouani, S. E. Pokossi, S. Mezghani, R. Vargiolu, H. Jacobs, J. Piezanowski, "Characterization of the Painted Surfaces Appearance by Continuous Wavelet Transform", 9th International Conference on Metrology and Properties of Engineering Surfaces, 10 – 11 Septembre, Halmstad University, Sweeden, pp.56-66, 2003.

[3] Mezghani, S., Zahouani, H., Piezanowski, J.-J., Multiscale characterizations of painted surface appearance by continuous wavelet transform, Journal of Materials Processing Technology 211 (2), 2011, pp. 205-211.

[4] H. Zahouani, S. Mezghani, R. Vargiolu, M. Dursapt, "Multi-scale study of abrasion signature by 2D wavelet decomposition", Wear, Vol. 264 (5-6), 2008, pp. 480-485.

[5] Tang, K.; Dimenna, R.; Buckius, R.: "Regions of validity of the geometric optics approximation for angular scattering from very rough surfaces", International Journal of Heat and Mass Transfer, Volume 40, Issue 1, pp. 49-59 (1997).

[6] Bergström, D.; Powell, J.; Kaplan, A.: "A ray-tracing analysis of the absorption of light by smooth and rough metal surfaces", Journal of Applied Physics, Volume 101, Issue 11, pp. 113504/1-11 (2007).

[7] Bergström, D.; Powell, J.; Kaplan, A.: "The absorption of light by rough metal surfaces - A three-dimensional ray-tracing analysis", Journal of Applied Physics, Volume 103, Issue 10, pp. 103515/1-12 (2008).

Journal of Physics: Conference Series **311** (2011) 012027 doi:10.1088/1742-6596/311/1/012027

Calibration of the z-axis for large-scale scanning white-light interferometers

S Boedecker[1], C Rembe[1], H Schmid[1], T Hageney[2], T Köhnlein[2]
[1]R&D Polytec GmbH, Polytec-Platz 1-7, 76137 Waldbronn, Germany
[2]eumetron GmbH, Gartenstraße 133, 73430 Aalen, Germany

E-mail: s.boedecker@polytec.de

Abstract. The calibration of white-light interferometers is an important requirement to make the results of these fast and accurate measurement tools comparable to the well-established tactile techniques. In this paper we present an approach for the calibration of the z-axis of a large-scale scanning white-light interferometer (SWLI) covering the full scanning range by utilizing a newly developed step-height calibration standard. The standard was specially adapted to the requirements of the large-scale WLI with an x-y-field of view of 38 x 28 mm^2 and a z-scan range of 70 mm. We have developed a new procedure to calibrate the z-scale by using the measured WLI data of the step heights of the standard to obtain an inverse characteristic to compensate non-linearities of the length reference, which is in our case simply the lead screw of the step-motor-driven linear stage. We describe the development of our standard, present measurement results before and after the calibration and discuss the uncertainties of the procedure.

1. Introduction

Scanning white-light interferometry [1][2] is gaining more and more importance in research and development as well as in manufacturing quality control. So far mainly classic microscopic white-light interferometers have been used to measure small object features [3]. However, the improvements in precision mechanical manufacturing have caused an increasing demand for large-scale high precision measurements [4][5] with uncertainties in the micrometer range. In comparison to microscopic setups, large-scale scanning white light interferometers use a Michelson setup of the interferometer, in which the object under test is imaged onto the camera through the beamsplitter by a telecentric optics. The reference path is fixed and the whole interferometer is moved in respect to the specimen by a translation stage. Interference appears always in the focus of the telecentric objective due to the optical operation principle and, thus, these kind of white-light interferometers can have a scanning range of several mm or even tens of mm. Therefore, motor driven linear stages have to be employed for the z-scan instead of high-precision piezos. In opposite to piezoelectric stages, motorized stages can exhibit a considerable measurement error from non-linearities in the motion. Consequently, for high-precision measurements the z-scale has to be calibrated. Although some publications [6][7] address this topic, until now no standard describes the procedure in detail.

Currently there is no detailed procedure which describes the measurement error of a large-scale WLI by means of an error budget. In addition the non-linearities of the linear stages used in the examined WLIs are random and can vary from device to device. At a first step we measured the non-linearities using a commercial high precision displacement laser interferometer. We tried to compensate the non-linearities of the linear stage by employing a lookup-table generated with the measurement data. However we could not see a significant improvement of the measurement error of our WLI. We concluded that the laser interferometer measurement is not directly relatable to the WLI measurement. In addition errors like cosine and Abbe error can falsify the result. So we had to find a different solution and followed a common method for the calibration of a measurement device by the comparison to traceable calibrated standards. In the case of dimensional measuring devices this could be for example high precision gauge blocks, or step height standards. For example coordinate measuring machines (CMMs) are commonly calibrated using a set of gauge blocks or a ball plate standard.

For the calibration of the z-axis of a WLI with a 70 mm scan range some special requirements have to be considered. Various step height standards are commercially available [8][9][10], but most of them only cover a height range from a few µm up to about 1 mm. Keeping in mind that the measurement range of large-scale WLIs exceeds 1 mm a potential step height standard for calibrating such WLIs must consist of multiple steps and cover the full z-range. Currently a depth setting standard with step heights up to a few mm is available from the Physikalisch Technische Bundesanstalt PTB [11]. This standard is not suited for calibrating the full z-scale of a large-scale WLI because it only has 6 different steps with a maximum height of 5 mm. Another approach is putting together a number of gauge blocks on a reference flat. But as the blocks are usually 9 mm wide only 4 blocks could be measured at once with the field of view (FoV) of the WLI.

Our approach addresses the calibration of the full z-scale by the use of a specially developed set of step height standards. With the help of these standards we were able for the first time to obtain a basic calibration of the z-axis of the large-scale WLI over the full z-range of 70 mm.

2. Development of the new Step Height Standards

2.1. Geometry

Our first idea to calibrate the z-axis by putting together gauge blocks of different heights on a reference flat was not practical. Commercially available blocks have a certain width and only 4 blocks could have been positioned in the 38 x 28 mm^2 FoV of the WLI. A calibration of the full z-axis of the WLI stitching together multiple measurements of gauge blocks would be too complex and time-consuming. In addition, a temperature variation over the necessary long measurement period could be a possible error source.

Another approach we had in mind was the manufacturing of customized gauge blocks with a width of only 1-2 mm to allow a higher number of steps within the field of view. Such blocks cannot be manufactured using hard metal as it would be necessary for the desired precision.

To solve these problems we have developed new step height standards to cover the full z-scale of the WLI with multiple steps in form of a stepped shaft, where every step is represented by an annular surface. Figure 1 shows a model of one of the step height standards. This geometry can routinely be manufactured with high precision in a cylindrical grinding process.

To define the exact geometry we had to consider the (1) special requirements of the measurement with our WLI, (2) limitations of the manufacturing process, and (3) the calibration procedure. First we had to define a reasonable number of steps which could be placed in the FoV. To ensure sharp edges in the manufacturing process and to account for the calibration procedure the steps could not be narrower than 1 mm. This still corresponds to about 7 to 8 pixel in width in the WLI measurement, which is appropriate for evaluation. Consequently, up to 19 steps of such a standard could be located in a field of view of 38 x 28 mm^2, measured simultaneously, and evaluated in an appropriate way. The steps were distributed equally over the whole z-range of the WLI (70 mm), which results in a distance of 3.75 mm. As with this coarse standard the z-scale can only be analyzed at some sample points, a potential non-linearity of the z-axis would not be fully captured. To solve this problem we developed a similar standard with a spacing of the steps of 0.295 mm (Figure 2). By merging the measurements of the fine and coarse standard the z-scale can be calibrated with the appropriate distance of the sample points. One of the annular surfaces of the steps was assigned to act as a reference plane (highlighted in Figure 1 and Figure 2). This assigned surface defines the basic coordinate system for the calibration of the standards and the WLI measurement. The origin of the the z-axis of the coordinate system is also defined by this reference plane. The assigned step was chosen because it corresponds to the position of the reference switch of the linear stage of the WLI. To assign the coordinate system for the calibration in x and y direction a flat was designed on one side of the standards (see Figure 1) acting as a coordinate reference. This flat also defines the lateral orientation of the standard for the WLI measurement.

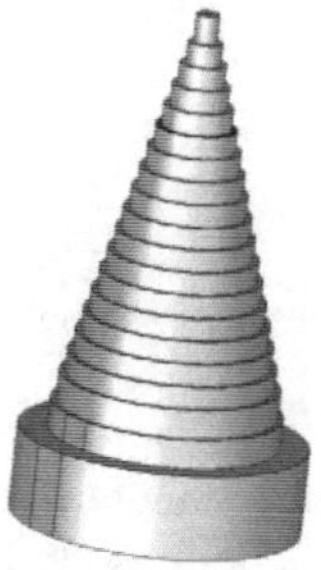

Figure 1. Model of the coarse step height standard.

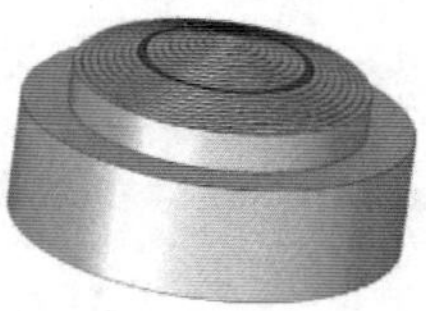

Figure 2. Model of the fine step height standard.

2.2. Manufacturing of the standards

We have searched for a suitable material that can be precision machined, which is durable and hard enough to withstand a calibration with stylus instruments. A low thermal expansion coefficient also was an important property. Considering the geometry of the standard, glass or other hard materials with a low thermal expansion coefficient have been discarded as material because of massive problems in the processing and the cost factor. So we have selected a special high-grade steel which is commonly used for precision gauges. The material (90MnCrV8) can be hardened and has a relative low thermal expansion coefficient of $\alpha = 12.2 \cdot 10^{-6}$ 1/K.

The standards finally were manufactured in a special cylindrical grinding process by [12]. To ensure a good surface quality with a low roughness, to account for the optical measurement method, the steps were additionally lapped. With this treatment we achieved a surface roughness of $R_z \sim 0.6$ µm. With this roughness the surface is nearly optical flat and, thus, the error due to speckle effects [13] is minimized. The flatness of the steps is below 2 µm, which is important for the data evaluation.

2.3. Calibration of the standards

Usually high precision stylus instruments are utilized to achieve a traceable calibration of standards for the micro and nano range, with the desired uncertainty. Stylus instruments could not be used for the calibration of our standards because these instruments usually have a vertical dynamic range of only a few millimeters, which is not sufficient for the high aspect ratio of our coarse standard. Consequently, a calibration with the desired uncertainty was only possible with a high precision CMM and appropriate care in the calibration process. The standards have been calibrated using a Carl Zeiss Prismo coordinate measuring machine. To achieve high comparability, the calibration procedure was adapted to the measuring and evaluation process for the calibration of the WLI. The evaluation area for every step was limited to eliminate errors due to the non-perfectly sharp edges of the steps. Every step was acquired as an annular surface with a width of 0.2 mm by measuring a number of radial profile lines. A plane was fitted to the single measurement points corresponding to each step. The distances D between the fitted planes and the reference plane (origin) on the standard were then calculated referred to the center of the standard to define the calibrated step heights (see Figure 3). The x-y plane and the origin of the basic coordinate system for the evaluation were assigned by measuring the reference plane on the standard. The flat on one side of the standard (see Figure 4) defines the x-z plane of the basic coordinate system when measuring with the CMM. The coordinate system during the calibration was defined to match the coordinate system of the WLI. The coordinate system is right handed where the tilt angles are defined as shown in Figure 4.

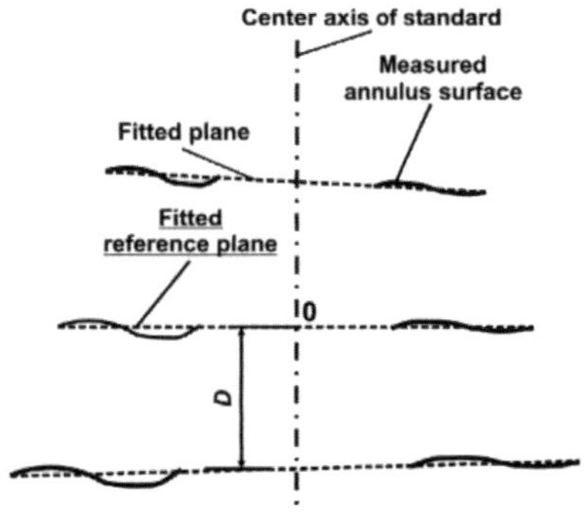

Figure 3. Calculation of the distances of the fitted planes to the reference plane.

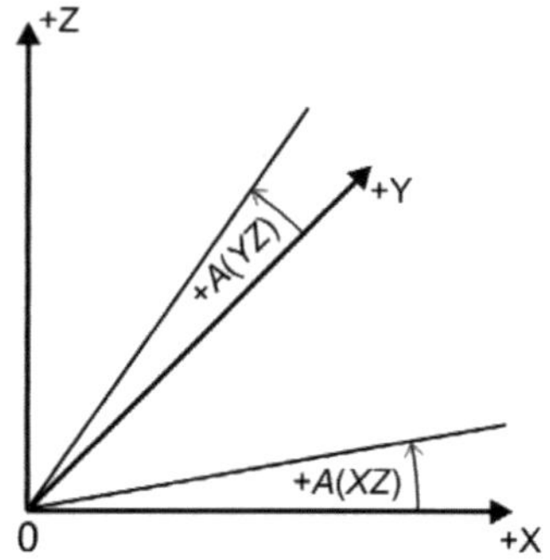

Figure 4. Definition of the coordinate system for the calibration of the standard and the WLI: *A*(*YZ*) - Angle in the YZ plane; *A*(*WZ*) - Angle in the XZ plane.

The following object features have been calibrated: The distance of the fitted plane for every annulus surface in respect to the reference plane on the standard (see Figure 3) and the angles in the XZ and YZ planes. Furthermore the flatness of the annulus surfaces of every step and the parallelism between the annulus surfaces and the reference plane according to ISO 1101 were calibrated.

To calculate the uncertainty a special procedure called virtual CMM which is described in [14] was applied. Taking into account the 1 mm width of the annulus surfaces at the steps and to keep the uncertainty as low as possible a stylus with 1 mm diameter was used for the measurement. To further reduce the uncertainty the substitution method [15] was used.

With the described calibration procedure we finally achieved an expanded uncertainty U ($k = 2$) for the step height/distance of $U = 0.4\mu\text{m} + 0.6 \cdot 10^{-6} \cdot D$, where D is the nominal distance in m to the reference plane on the standard. The maximum expanded uncertainty for the distance/step height for the calibration of our standard results in about 450 nm.

3. Calibration of the Large-Scale WLI

The first step of our calibration procedure for the z-axis is the measurement of the coarse standard. For this, the standard is positioned in the center of the FoV. The height of the standard is adjusted so that the interferences appear on the reference plane at the zero position of the z linear stage (Figure 5). Tip and tilt are adjusted to align the reference parallel to the measurement devices. The measured topography data (Figure 6) is evaluated with a software algorithm. Therefore, (1) the center of the topography of the standard is determined by means of image processing, (2) masks concentric to this center are set (Figure 6) to define the analysis area for every annulus surface of the steps with a width of 0.2 mm, and (3) the regression plane is calculated for the topography in every analysis area. To make the measurement comparable to the tactile calibration of the standard a low pass filter for the topography data is applied. Based on the evaluated angles of the regression planes and the calibrated angles of the annular surfaces the overall tilt of the standard is determined and corrected. Subsequently, the distances D of the annulus surfaces to the reference plane of the standard are calculated in the same way (Figure 3) as described in the section 2.3. The calibrated step heights are corrected by the measured temperature of the standard to derive a coarse deviation curve of the z-scale at certain sample points.

The size of the field of view and the necessary width of the annular surfaces of the standards limit the number of possible step heights of a single standard to approximately 20. Thus, it is necessary to employ an additional finer standard with a smaller step size. The Nyquist-Shannon Sampling Theorem [16] has to be considered to measure all relevant spatial frequency components of the z-scale non-linearities. Thus, a fine standard is measured in the same way, in a height position where its reference plane matches with one of the steps of the coarse standard (Figure 5) to cover the distance between the steps of the coarse measurement. Next the data is evaluated as described above to derive a finely sampled deviation curve for a small section of the z-axis.

In a post processing step the data of the two measurements is merged to achieve information of the z-scale with a finer spacing for the sample points. Therefore, the interpolated data of the fine standard is shifted by an offset so that the deviation curves for both measurements overlap in a certain region. The result is then used to generate a correction table (actual value/target value) for the z-axis which can be saved in the software of the device to calibrate the WLI. The values between the sample points are interpolated linearly.

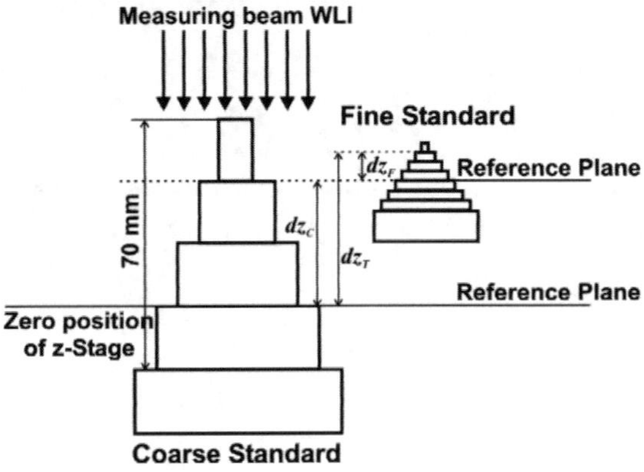

Figure 5. Schematic arrangement for the calibration procedure of the large-scale WLI with the help of the novel step height standards: dz_C - height difference for coarse standard; dz_F - height difference of fine standard; dz_T - Total height difference for sample point.

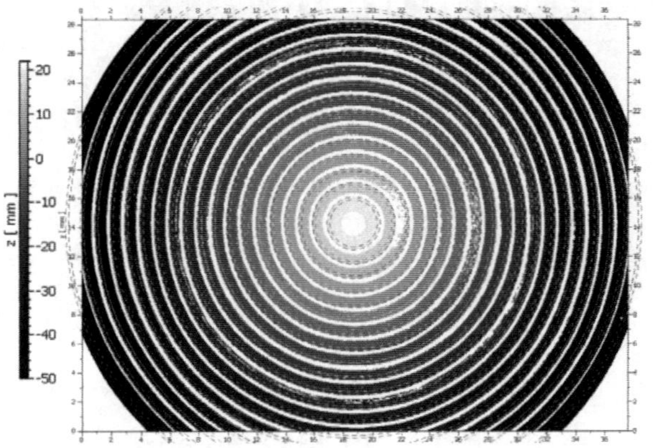

Figure 6. 2D color coded measured height result for the coarse standard (also shown are the analysis areas).

4. Discussion of the Error Sources

Currently, there still are some potential error sources in our calibration procedure that until now have not been analyzed in detail.

From previous studies we know that the linear stage of the examined WLI can have a considerable periodic position error which correlates with the lead screw pitch. As in our case the length reference is represented by the lead screw this error will contribute to the error of the WLI measurement. We have developed the fine standard that was used for these studies for various calibration purposes. Consequently, the spacing of the steps is not small enough to fully capture the periodic errors of the linear stage in respect to the Nyquist-Shannon-Sampling Theorem. As in the correction table for the z-axis calibration the values between the sampling points are interpolated linearly, with the standards we used it is not possible to calibrate all position errors of the linear stage.

There are also some error sources in the evaluation procedure for the measured WLI data. During the calibration of the standards with the CMM all annulus surfaces of the steps could be acquired completely. Due to the limited field of view, some of the steps can only be measured partially with the WLI (Figure 6). Thus, fitting a regression plane to an annulus which does not fit completely into the FoV is susceptible to flatness deviations and can cause an error in the measured distance *D*.

The merging of the measurement data from the coarse and fine standard could also cause an error when generating the correction table for the calibration of the z-axis. The individually measured distances with the coarse standard dz_C (Figure 5) exhibit an uncertainty. The measurements with the fine standard dz_F also have an uncertainty. Consequently, when merging both measurements the uncertainty for dz_T is approximately the sum of the two contributions.

As the standards were made from a high-grade steel the thermal expansion coefficient is not negligible for step sizes of a couple of tens of mm. The temperature of the standards is acquired during the measurement to account for a temperature difference between the standards and the surrounding. An accurate measurement of the standard's temperature is not a simple task. A measurement error of the temperature of the standards during the data acquisition can falsify the correction of the calibrated distances of the steps in the evaluation procedure. Therefore, an uncertainty in the temperature measurement results in a step height error, which also contributes to the overall error of the calibration procedure.

Another question is the long-term stability of the calibration. Due to wear and aging of the linear stage the linearity behaviour could change over time. At the current point we cannot make a statement about the long-term stability of the calibration procedure. The investigation of these uncertainty contributions will be addressed in our future studies.

5. Example Calibration

Figure 7 shows an example result for the calibration of a large-scale WLI. From the comparison of the curves it can clearly be seen that the measurement deviation of the WLI on average can be reduced by a factor of 3 when employing the newly developed step height standards and our calibration procedure.

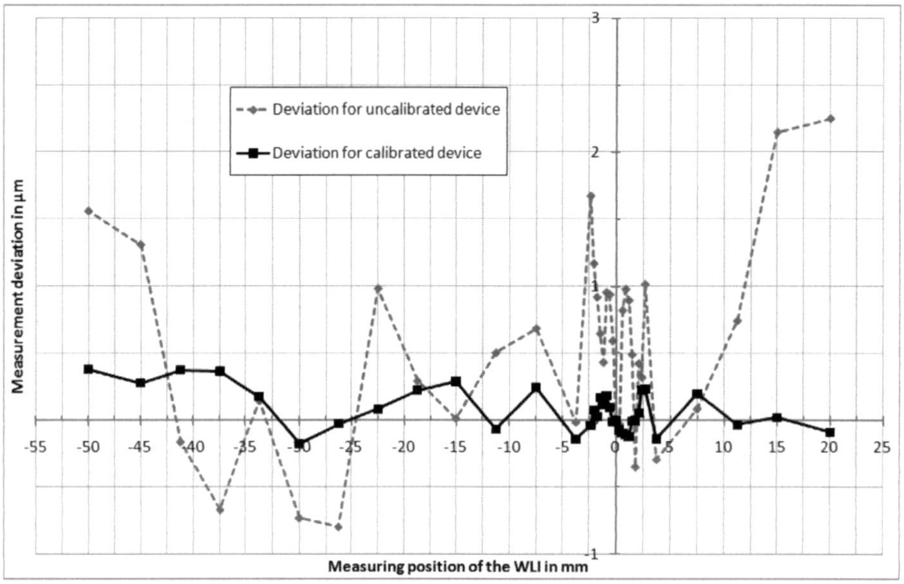

Figure 7. Comparison of the deviation curves for an example calibration of a large-scale WLI.

6. Conclusion and Outlook

We have described the development of two new step height standards which enable a detailed calibration of the full z-scale of a large-scale Scanning White-Light interferometer with a vertical scanning range of 70 mm. We have designed two standards in form of a stepped shaft (Figure 1, Figure 2), where every step is defined by an annular surface. The coarse standard has 19 steps with a spacing of 3.75 mm. The fine standard consists of 18 steps with a spacing of 0.295 mm. Every standard has a reference plane to trace the calibrated sample distances to the zero position of the motorized linear stage of the WLI.

In addition, we have discussed the special requirements for the standards in respect to measurements with a large-scale WLI with a field of view of 38 x 28 mm^2. The requirements for a traceable calibration of the standards have also been pointed out. With a special calibration procedure we have achieved a maximum expanded uncertainty for the calibrated step heights of about 450 nm.
The new standards make possible for the first time a new approach for the calibration of the full z-axis of large-scale WLIs. We developed a procedure to calibrate the full z-axis of a large-scale WLI with the help of our novel standards. The calibration procedure can account for all possible non-linearity errors in the z-scale by merging the measurements of a coarse and a fine standard. Reference planes are utilized to realize the combination of the samples from multiple measurements. The result of an example calibration for a large-scale WLI (Figure 7) proofs that the measurement error can be reduced on average by a factor of 3 with our procedure.
Future work will address a detailed analysis of the error sources and will employ a fine standard with a step spacing small enough to sample all spatial frequency components of the linear stage non-linearity according to the Nyquist-Shannon Theorem.

Acknowledgments

This work is supported by the German Federal Ministry of Economics and Technology (BMWi) grant number 01FS10046.

References

[1] Byron S. Lee, Timothy C. Strand 1990 Profilometry with a coherence scanning microscope, *Appl. Opt.* vol **29** No. 26
[2] Dresel T; Häusler G and Venzke H 1992 Three-Dimensional sensing of rough surfaces by coherence radar *Appl. Opt.* vol **31** No. 7
[3] Zecchino M 2004 Quantifying Laser-Induced Thermal Deformation of a MEMS device, *Semiconductor Manufacturing* vol **5** Issue 8
[4] Polytec GmbH, Application Note: 3-D Measurements on Shock Absorber Pistons, http://www.polytec.com/int/products/surface-metrology/tms-100-topmap-metrolab/
[5] Bauer W 2007 Praxis von 3D-Messungen mit Nanometergenauigkeit von Strukturen, Maß, Form und Lage –im Labor und in der Linie *VDI-Berichte Optische Messung technischer Oberflächen* No. **1996** pp. 253-262
[6] VDI/VDE Guideline 2655 Part 1.3, currently under preparation
[7] Boedecker S et al. 2010 Comparability and uncertainty of shape measurements with white-light interferometers *Proc. SPIE* 7718-18
[8] HALLE Präzisions-Kalibriernormale GmbH, KNT 4080/03 depth setting standards, http://www.halle-normale.de/
[9] SiMETRICS GmbH, VS vertical standards, http://www.simetrics.de/en/
[10] VLSI Standards, Incorporated, SHS step height standards and UTSHS step height standards, http://www.vlsistandards.com/
[11] Brand, U, Schnädelbach H, Schödel R, Feist C and Hinzmann G 2006 New depth-setting standards with grooves up to 5 mm depth *Proc. Euspen* pp 438-441
[12] Blankenhorn GmbH, Ruiter Straße 51, 73734 Esslingen, Germany
[13] Pavliček P and Hýbl O 2008 White-light interferometry on rough surfaces-measurement uncertainty caused by surface roughness *Appl. Optics* **47** pp 2941-2949
[14] VDI/VDE Guideline 2617 Part 7, Estimation of measurement uncertainty of coordinate measuring machines by means of simulation
[15] DIN ISO/TS 15530 Part 3, Geometrical product specifications (GPS) - Coordinate measuring machines (CMM): Technique for determining the uncertainty of measurement - Part 3: Use of calibrated workpieces or standards
[16] Shannon C *1949* Communication in the presence of noise *Proc. of the IRE* vol **37** 10-21

Confocal Line Scanning Sensor

S. Chanbai,[1] I. Zafarullah,[2] G. Wiora,[1] L. Wewer[1] and H. Roth[3]

[1]NanoFocus AG, Lindnerstr. 98, 46149 Oberhausen, Germany
[2]Applied Scientific Imaging Inc., Toronto, Canada
[3]Institute of Automatic Control Engineering, University of Siegen, Hölderlinstr. 3, 57068 Siegen, Germany

E-mail: [chanbai|wiora]@nanofocus.de

Abstract. We have developed a novel confocal-based imaging sensor for surface characterization. In this case, a tilted-plane technique is incorporated in a confocal imaging system to create a new parallel scanning scheme, enabling the sensor to be designed and developed as a robust and simple configuration. With a tilted disk consisting of in-line pinholes, a motionless parallel z scanning scheme is manifested when the specimen is transversely scanned through the stationary diffraction-foci projecting at different depths. This sensor uses a line scanning approach, so that it is entitled as a *Confocal Line Scanning Sensor* (CLSS). In this paper, the CLSS principle, the concept of data processing, and major calibration are described. The sensor was first developed as a two-dimensional profiler to cover the measurement ranges of up to 50 μm in depth and up to 15 mm in lateral length. Experimental results were carried out using calibrated specimens for roughness measurement. In this system, the optical lateral resolution is 0.5 μm, and the depth resolution, defined by noise-limited approach, is 15 nm.

1. Introduction

Increasing the speed of measurement and reducing the complexity of the setup pose significant challenges in the development of a confocal imaging system for price reduction and new market penetration. The confocal imaging technique, first described by Minsky in 1961 [1], has become a powerful tool for surface characterization. Using a pinhole to function as a spatial filter, a confocal imaging system can perform the optical sectioning also known as the depth-discrimination property, and enhance the lateral resolution as compared to a conventional light microscope [2]. However, these enhancements in resolution and in optical sectioning are substituted by the need to scan an observed focal-point in three dimensions to render a 3D surface topography. A real-time Nipkow disk, first incorporated in a confocal microscope by Petrain and Hadravsky [3], can facilitate only the raster scan in the lateral plane ($x - y$), in which case the depth scanning is still required.

In such a circumstance, combining the techniques of motionless parallel depth (z) scanning and lateral scanning become a single solution to increasing the speed of a large area measurement. In this paper, a novel technique of motionless parallel depth scanning is demonstrated using a tilted-plane technique [4]. Pursuant to that, a feasible simple and robust setup was designed and developed to preserve the high resolution of a confocal imaging system and the high light-budget efficiency per wavelength of quasi-monochromatic illumination.

Journal of Physics: Conference Series **311** (2011) 012028 doi:10.1088/1742-6596/311/1/012028

1.1. Principle of the Confocal Line Scanning Sensor (CLSS)

The CLSS inspects the specimens using a back-reflection scheme; the emitted light is reflected backward from the specimen to the CCD. In this new approach, a tilted pinhole mask is used to generate multiple optical probes or diffraction-foci projecting at different depths with uniform spacing ($\Delta x, \Delta z$) in the tilted focal plane, as shown in Figure 1(a). Following the Scheimpflug's rule [5], the measurement range in depth can be extended by increasing the tilt angle of the pinhole mask and so the focal plane. When the specimen is transversely scanned through these foci, a parallel depth scanning is manifested in the $z-$axis, achieving 160 sampled points per shot. In consequence, the transverse scan of the specimen enables a measurement of a two-dimensional profile of infinite length. Moreover, the stationary diffraction-foci are formed using an optical setup comprising no moving parts, making the setup robust and simple.

1.2. Experimental setup

Key components of the CLSS are presented in Figure 1(b). A light emitting diode (LED), emitting green light at $\lambda = 532$ nm wavelength, is exploited as a stable, inexpensive, and incoherent light source. To capture the reflected light signal, a charge-coupled-device (CCD) camera (AVT Dolphin, F145B) is employed. In this design, the Köhler illumination is incorporated to maintain the most uniform illumination, realizing with condenser and collector lenses. An infinity-corrected objective applied in the system helps to facilitate a straightforward calculation in the pinhole mask design. Therefore, an infinity-corrected 20× objective and a matching tube lens (f = 180 mm) from Olympus is selected to carry out the experiment. To eliminate the back reflection from the pinhole mask, the beam from the beam splitter to the CCD is tilted with respect to the normal axis. This method helps minimize the number of optical surfaces used in the setup, as compared to applying polarizers and a quarter-wave plate. The measurement is then conducted using a stepper motor (Haydon 35000 Series, E35H4N-05) to scan the specimen in the lateral direction at 5.8 frames per second (fps).

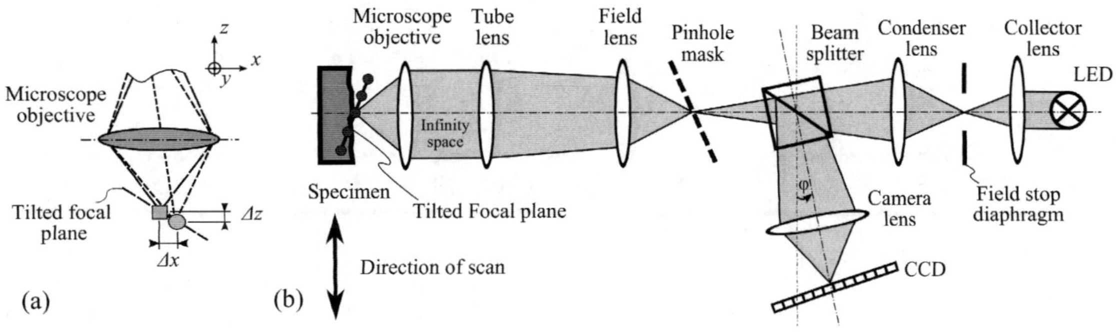

Figure 1. CLSS setup. (a) Tilted focal plane. (b) Schematic of the experimental setup.

2. Data Processing and Calibration

2.1. Concept of data processing

A significant role of the data processing is extracting the surface height information from the intensity output. The maximum reflected light intensity carries the surface height information in the form of a confocal curve, as shown in Figure 2(c). In consequence, new algorithms for data processing were developed particularly for use in the CLSS. In the CLSS, the data is acquired in the order that is delineated in Figure 2(a). The data set is then sorted to meet the same $x-$coordinate as demonstrated in Figure 2(b). When the intensity along the z axis is plotted

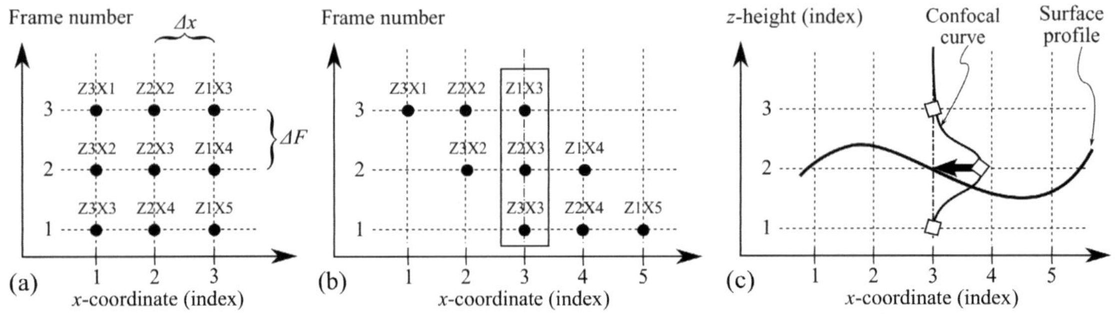

Figure 2. Concept of the data processing in the CLSS [6]. (a) An acquired data stack. (b) A sorted data stack. (c) An intensity distribution curve formed at a given $x-$coordinate, known as a confocal curve. A confocal curve carries the surface height information as a function of the maximum intensity.

at a given $x-$coordinate, the confocal curve is realized and then analyzed to reveal the surface height, based on peak finding approaches [7,8]. Finally, a two-dimensional surface profile is reconstructed in the $(x - z)$ plane.

2.2. System Qualification and Calibration

The qualification of the system parameters is needed to run the data reconstruction. The central parameter is the ratio (r) of the lateral scanning acquisition step size (ΔF) to the lateral focal interval (Δx): $r = \Delta F/\Delta x$.

This parameter is used to sort the data as shown in Figure 2(b). To reveal r, a lateral displacement standard with periodic chrome stripes, approximately 100 nm high, on glass is used. It is placed so that the stripes are perpendicular to the scanning direction, and an image stack as shown in Figure 3(a) is acquired. The parallel stripes in this image stack have an inclination defined by r. The inclination r is measured using edge detection modules contained in computer-aided vision systems.

With the knowledge of r, the scanning axis step size can then be calibrated. From the measurement in Figure 3(a), the confocal curve is reconstructed as shown in Figure 2(c). For each x position of the scanner, the maximum reflected intensity is measured and the periodic pattern is plotted over the uncalibrated x position, as shown in Figure 3(b). The autocorrelation of the periodic pattern is calculated and its maxima and minima locations are plotted as shown in Figure 3(b). By calculating the mean period length ΔP of the maxima and minima along the $x-$axis, the calibrated $x-$coordinate can be computed from the uncalibrated coordinate (x') using the absolute period length (τ): $x = x' \cdot \tau/\Delta P$.

Finally the axial focal interval (Δz) is calibrated by a standard procedure measuring a groove normal as described in the ISO 25178-701 standard [9].

3. Experimental Results and Discussion

The experiments were carried out on several calibrated specimens, using 20 × objectives. The results in Figure 4 demonstrate that roughness measurement is feasible with the CLSS. In consequence, we compared the CLSS measurements in Figure 4(a) to a commercial confocal microscope, μsurf custom from NanoFocus, in Figure 4(b) using the very same objective. For the CLSS, a good agreement of the surface roughness value R_a = 1.66 μm to the calibrated value of the standard ($R_a = 1.63 \pm 0.03$ μm) could be found. In Figure 4(b), the μsurf reveals

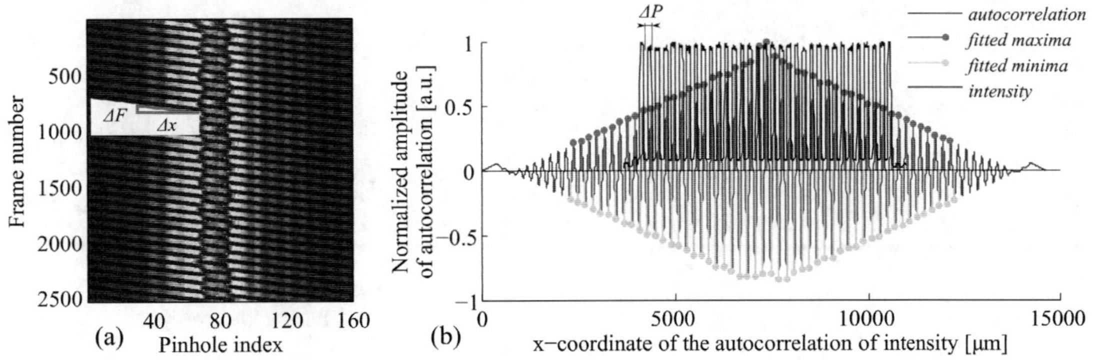

Figure 3. Calibration of the lateral focal interval (Δx). (a) Measured data of a square periodic pattern. (b) Autocorrelation of the measured data of a square periodic pattern.

more details of the profile in accordance with a higher sampling density using a finer lateral-sampling interval of 0.8 μm, as compared to a relatively large lateral-sampling interval of 2.5 μm used in the CLSS. In addition, the interpolation of the raw data, applied in the data sorting in Figure 2(b), may further smooth the confocal curve in Figure 2(c), which will filter the high frequency components of the reconstructed profile as the end effect.

On the other hand, the new parallel z scanning scheme of the CLSS has a distinct advantage in adjusting the data density, to improve the resolution in the lateral direction. However, the higher data-sampling density, the more measurement time are needed and the lower the measurement speed. Basically, the lateral resolution of the CLSS is a function of the sampling interval in the lateral direction, which follows the Nyquist sampling theorem, and limited by the optical lateral resolution. The optical lateral resolution of the CLSS is 0.5 μm, based on 20× 0.45NA objectives and $\lambda = 532$ nm, where the optical lateral resolution is the FWHM of the lateral intensity response. In contrary, the axial resolution is defined using noise limited approach. The noise limited precision can be verified straight forward by referring to the root-mean-square R_q of the profile measurement on a flatness standard [10]. Accordingly, the noise-limited axial resolution of 15 nm was obtained with the CLSS, using a mirror of $\lambda/4$ flatness.

Figure 5 depicts the axial intensity response obtained with a flat mirror, known as a confocal curve. The graph shows that the FWHM is 3.5 μm, which is 1.52 times larger than the theoretical value of 2.3 μm. This is partially caused by using pinholes larger than the theoretical optimum.

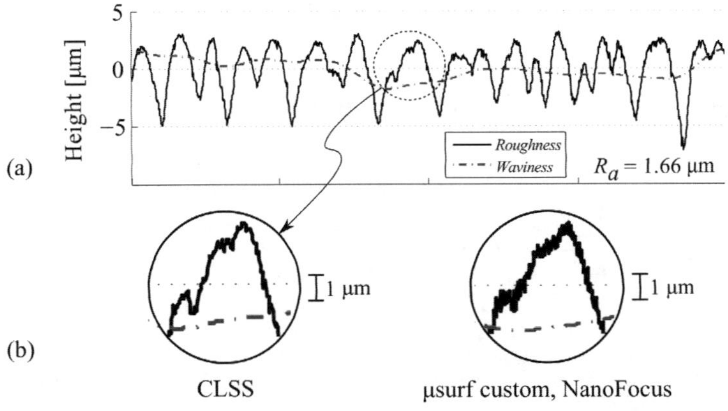

Figure 4. Measurement results of a roughness standard calibrated by Physikalisch Technische Bundesanstalt (PTB) of $R_q = 1.63$ μm. (a) A profile obtained with the CLSS, $R_a = 1.66$ μm. (b) Details of the same place compared to a commercial confocal system, μsurf custom from NanoFocus.

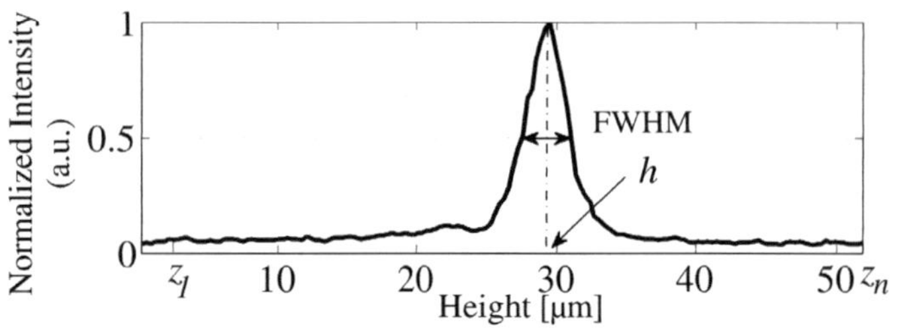

Figure 5. The axial intensity response which is termed as a *confocal curve* obtained on a flat mirror.

By evaluating the maximum peak of the confocal curve, the surface height (h) at a given x coordinate is calculated using the centre-of-mass algorithm [7,8] of the form

$$h = \frac{\sum_{z_p \in FWHM} I_{z_p} \cdot z_p}{\sum_{z_p \in FWHM} I_{z_p}},$$

where I denotes the intensity output obtained in the axial coordinate (z_p), and $p = 1, 2, ..., n$ are the indices of data sampled in the axial/vertical direction.

4. Conclusion

In conclusion, a novel parallel depth scanning scheme of the CLSS offering a new solution for two-dimensional (2D) optical profiling with a simplified and robust configuration is verified in this paper. Based on this new measuring scheme, a tilted-plane technique is originally incorporated in a confocal imaging system. Accordingly, the algorithms for data processing and calibrations were particularly developed to support the new measuring approach. The experimental results demonstrate the feasibility of measuring surface roughness by using the CLSS, attaining the optical lateral-resolution of 0.5 μm and the noise-limited axial resolution of 15 nm. Nonetheless, the CLSS can be further developed as a three-dimensional surface topographer, when the variation of the magnification introduced by the tilt is corrected. In the current setup, the CLSS can cover the measurement ranges of 50 μm in depth and of 15 mm in lateral length, with the measurement speed of 5.8 fps. However, the measurement ranges can be enhanced using a high NA objective at low magnification, whereas the speed of measurement can be further increased using a faster data acquisition technique and a faster camera.

5. Acknowledgements

This work was supported by NanoFocus AG and the Marie Curie Research Training Network, European Community's Sixth Framework Programme, under the EC contract MRTN-CT-2006-035589.

References

[1] Minsky M 1961 (US Patent 3013467)
[2] Wilson T 1990 *Confocal Microscopy* (Academic Press)
[3] Petrain M and HadravskySze M 1969 *J. Opt. Soc. Am.* **58** 661
[4] Chanbai S, Valentin J, Weber M and Wiora G 2010 (German Patent DE102008031412 A1)
[5] Scheimpflug T 1904 (British Patent GB 1196/1904)
[6] Chanbai S, Wiora G, and Weber M 2009 *Proc. SPIE* **7378** 737822
[7] Jordan H J, Wegner M and Tiziani H 1998 *Meas. Sci. Technol.* **9** 1142
[8] Ruprecht A K, Wiesendanger T F and Tiziani H 2002 *Appl. Opt.* **41** 7410
[9] ISO 25178-701, Geometrical product specifications (GPS) - Surface texture: Areal - Part 701: Calibration and measurement standards for contact (stylus) instruments
[10] ISO 4287, Geometrical Product Specifications (GPS) - Surface texture: Profile method - Terms, definitions and surface texture parameters

An iterative algorithm for calculating stylus radius unambiguously

T V Vorburger, A Zheng, T B Renegar, J-F Song, and L Ma

National Institute of Standards and Technology, 100 Bureau Drive, Gaithersburg, MD 20899, USA

Email: tvtv@nist.gov

Abstract. The stylus radius is an important specification for stylus instruments and is commonly provided by instrument manufacturers. However, it is difficult to measure the stylus radius unambiguously. Accurate profiles of the stylus tip may be obtained by profiling over an object sharper than itself, such as a razor blade. However, the stylus profile thus obtained is a partial arc, and unless the shape of the stylus tip is a perfect sphere or circle, the effective value of the radius depends on the length of the tip profile over which the radius is determined. We have developed an iterative, least squares algorithm aimed to determine the effective least squares stylus radius unambiguously. So far, the algorithm converges to reasonable results for the least squares stylus radius. We suggest that the algorithm be considered for adoption in documentary standards describing the properties of stylus instruments.

1. Introduction

In stylus measurements of surface roughness, the finest topographic details that can be measured on the surface depend on the shape and size of the stylus tip. Hence, the stylus radius is an important specification for stylus instruments. Recommended values of stylus radius are given in national and international documentary standards and values are commonly provided by instrument manufacturers. However, it is difficult for users or manufacturers to measure the stylus radius unambiguously. Accurate profiles of the stylus tip may be obtained by profiling over an object sharper than itself, such as a razor blade [1-5] (see figure 1), or by scanning electron microscopy (SEM) or by optical microscopy for large tips. However, the stylus profile thus obtained is a partial arc, spanning in most cases ±45°, and unless the shape of the tip is a perfect sphere or circle, the effective value of the radius depends on the length of the tip profile over which the radius is determined.

We have developed an iterative, least squares algorithm aimed to determine the effective stylus radius unambiguously. The procedure starts with an initial estimate of the radius, such as with a specification provided by the manufacturer, and performs a least squares fit over a profile length consistent with the manufacturer's specification. If a different least squares radius is obtained, then a new profile-length parameter is calculated and the least squares procedure is applied again until convergence is reached. We have tested the algorithm so far with vertical least squares circle fits of measured stylus profiles, which tend to be flatter on top than near the flanks. So far, the algorithm converges to reasonable results for the least squares stylus radius. We suggest that the algorithm be considered for adoption in documentary standards describing the properties of stylus instruments.

Journal of Physics: Conference Series **311** (2011) 012029 doi:10.1088/1742-6596/311/1/012029

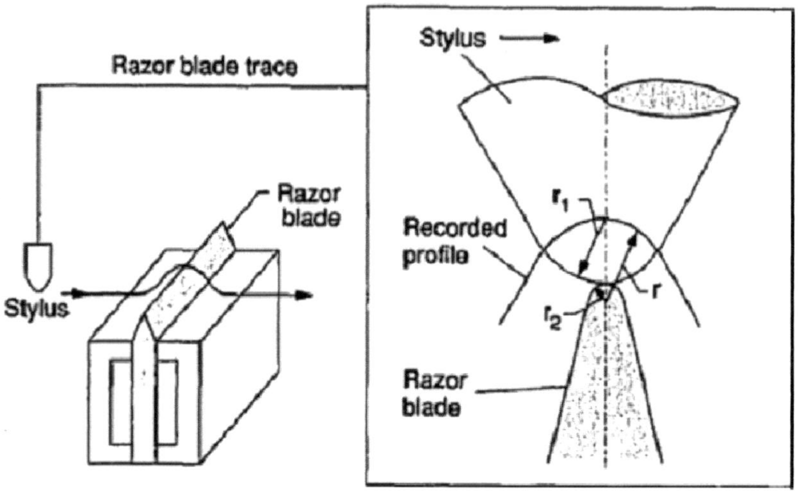

Figure 1. Schematic diagram of the razor blade trace method for profiling a stylus tip [2,4,5].

1.1. Background

Determining the effective least squares radius of a stylus tip involves two steps: measuring the stylus tip profile accurately and extracting an effective radius from the measured tip profile. There are several ways to measure stylus tip profiles with sufficient accuracy [1]: SEM, optical microscopy or optical projection, and scanning the stylus tip over a sharp object such as a razor blade [1-5]. SEM has high resolution and is certainly useful for stylus tips with radii as small as several nanometers. Optical microscopy and optical projection have lower resolution but are useful for determination of stylus radii of about 10 μm and perhaps somewhat smaller. The razor blade approach has been written into documentary standards [4,5] and has been shown to be useful for profiling stylus tips with radii as small as about 0.1 μm [2] or as large as about 12 μm [1].

Once the profile has been measured, the second step of determining the effective radius from a measured stylus tip profile is prone to ambiguity. Due to imperfections in the shape of the stylus tip profile, the computed value of the radius depends on the length of arc or area of tip surface over which the radius is calculated. Further, for any measurement application on real engineering surfaces the effect of the stylus radius depends on the area of contact between the tip of the stylus and the surface and hence on the local topography of the surface. The functional stylus radius, which is related to the outside envelope of the stylus with the contacted surface, is therefore, dependent on the particular application. For the present paper, we ignore the latter problem and confine ourselves to determining the radius of the stylus unambiguously in a standard way.

1.2. Focus

We also limit our focus in other ways. The aim is to enable vendors and users of stylus instruments for surface profiling to agree on a quantitative measure of the quality of a stylus tip, independent of any particular application. First, for simplicity, we limit ourselves to 2D profiling and the determination of an effective circular radius because the results are simpler to describe and visualize. However, the method can be extended to 3D topography measurements and to determining effective spherical radius from 3D data. Second, we limit the discussion to standard least squares fitting of the stylus tip profile to the parameters of a circle. Whereas from a functional point of view, it would be preferable to construct an effective stylus profile based on its points of contact rather than its mean profile, this brings us back to issues regarding the shape of the measured surface and to specific applications. Third, for simplicity, we limit the least squares fitting process here to minimizing the residuals in the vertical direction rather than the radial direction.

Journal of Physics: Conference Series **311** (2011) 012029 doi:10.1088/1742-6596/311/1/012029

Figure 2. Razor blade profile of a stylus that was in use for many years in our laboratory. The stylus was chisel shaped with a flat about 4 μm across at the tip [1].

Even with this narrow focus, the least squares fitting of the partial arc of a stylus tip profile is ambiguous because the fitted radius depends on the length of profile used for the fitting process, unless the measured tip profile is perfectly round. We illustrate the issue with the profile of a fine stylus used for a number of years in our laboratory. This stylus (figure 2) has a flat across the tip about 4 μm long. It is certainly useful for probing peak or valley surface features more than 4 μm across but is virtually useless for discerning features less than about 4 μm across. How does one determine the effective radius of the stylus from a profile of its projected shape? The issue may be highlighted more starkly by considering the extreme case of a simulated, perfectly flat tip shown later on the left hand side of figure 4. The effective radius is infinite for closely spaced features but is approximately 4 μm for features that are spaced more widely than, say, 5 μm. Accordingly, a least squares fit of the central 4 μm of the tip profile or smaller should yield an effective radius of infinity, but a fit of length 5 μm yields a finite radius.

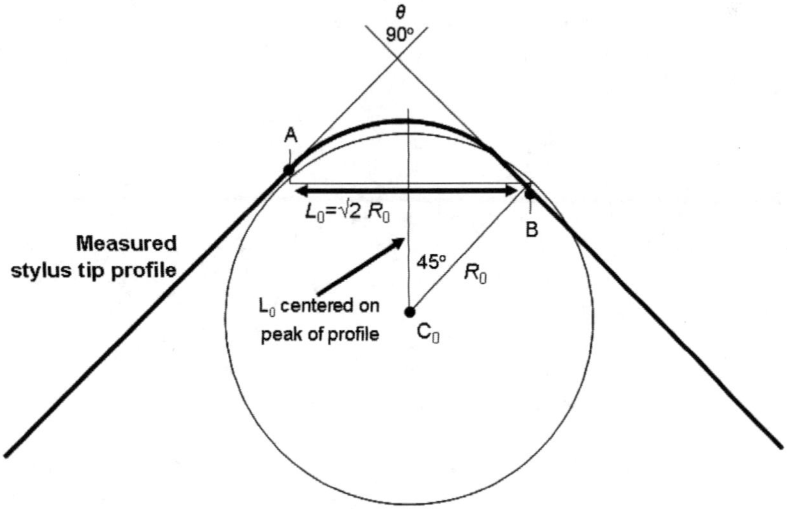

Figure 3. Construction of the parameters of a circle for the first iteration of the least squares fitting routine to determine the radius of the stylus tip from its measured profile. The initial guess for the radius is R_0; the chord length L_0 is ($\sqrt{2}$) R_0 for a tip with 90° apex angle. This chord length defines the profile to be fitted and is centered about the highest point on the profile. The center of this initial circle is C_0, and A and B represent the extremes of the measured profile to be fitted initially.

2. Stylus Tip Fitting Algorithm

The algorithm is an iterative least squares fitting process that works as follows:

- The profile of the stylus tip is measured by one of the methods described above.
- An estimate (R_0) is made of the stylus tip radius, based on the manufacturer's specification or on a known value from a previous measurement.
- The cone apex angle θ is calculated from the straight shanks of the measured stylus profile or is known beforehand.
- Using the values of R_0 and θ, the horizontal chord length (L_0) of the ideal circular stylus tip profile is calculated, that is, the horizontal distance between the transition points where the circular profile would meet a straight shank with apex angle θ. For a stylus tip with a 90° apex angle, L_0 is equal to $(\sqrt{2})R_0$. This is illustrated in figure 3. For arbitrary θ and R_0, the formula is

$$L_0 = 2\ R_0 \sin\,(90° - \theta/2) \quad (1)$$

- The length L_0 serves as the horizontal length of the stylus tip profile to be fitted. This chord length is then centered on the highest point in the measured stylus tip profile and a least squares fit of the associated segment of profile is performed to the parameters of a circle. The fitted segment of profile may include portions of the shank as shown here by the profile portion near B in figure 3.
- The parameters of the fitted circle then determine new values for the radius R_1 and center of the circular profile (x_1, y_1). The fitted radius R_1 is used to calculate a new segment L_1, now given by $2\ R_1 \sin(90° - \theta/2)$, which is centered on the x-coordinate (x_1) of the fitted circle and a new least squares fit is performed.
- The values of R, L, and x are iterated until the calculation converges to a point where the successive values for R are sufficiently close together. Our current criterion is a difference of 0.01 µm.
- A damping fraction for the changes in R, L, and/or x can be chosen so that the calculation does not become unstable. That is, the full correction resulting from each least squares fit is not applied with each iteration but only a fraction of it.

3. Results

Results are given for two cases: the simulated flat tip stylus and the measured tip profile of a stylus currently used in our laboratory for calibrations. The left hand side of figure 4 shows the results for the flat tip having a width 4 µm across. The apex angle is 90°. A slight tilt was applied to this profile so that the highest point of the profile was at the right hand corner of the chisel shape. Thus the fitted circles were made to start well off center. Nevertheless, the routine converged in approximately five iterations to a result for R of 4.7149 µm from a starting estimated value of 5 µm. From a different starting value of 2 µm, the routine converged in about eight iterations to the same final value of 4.7149 µm for R.

The right hand side of Fig. 4 depicts a razor blade profile for a stylus tip currently used in our laboratory. The stylus has a nominal apex angle of 90° and a nominal radius of 2 µm. The least squares fitting routine converges in about seven iterations to a radius of 1.4455 µm. This type of result, where the fitted radius is smaller than the nominal radius assigned by the manufacturer, is consistent with other results we have obtained for similar types of tips. The combined standard uncertainty that we estimate for the above result is 0.11 µm. It depends on the Type A repeatability, the uncertainty of the x- and z- magnification, the size of the razor blade tip, which is about 30 nm, and on a model we use to estimate the effective radius for different types of surfaces, where we calculate the least squares radius for a length of tip profile extending over ± 15°.

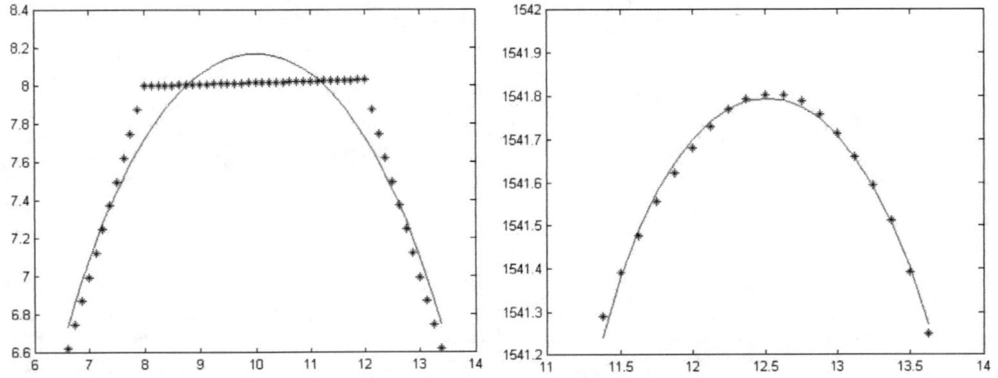

Figure 4. All units are in μm. (Left) data points simulating a stylus tip profile with a flat 4 μm across and curve showing the least squares fitted profile resulting from the fitting algorithm discussed here; the fitted radius is 4.7149 μm. (Right) measured razor blade profile (data points) of a stylus tip with a nominal radius of 2 μm and the resulting fitted profile; the fitted radius is 1.4455 μm. Note that the aspect ratios of the graphs are not one-to-one. The standard uncertainties of individual data points in the right hand profile are approximately 1 nm in the vertical direction and 0.02 μm in the horizontal direction.

4. Discussion

The goal of this procedure is not to determine the effective radius of a stylus tip for profiling surfaces with any surface application in mind; the number of applications of surface finish measurement is too large for us to be able to provide a procedure that is optimally suitable for all users. Rather, the goal is to provide a common, unambiguous approach to calculate stylus radius from a partial arc of say, 90° or 120°, that vendors and users can agree to. The procedure could then be designated in documentary standards. If required, standards could further designate acceptable deviations of the stylus tip profile from the fitted radius. Currently, the International Organization for Standardization (ISO) standards for surface texture measurement do not provide a procedure to calculate stylus radius, and the related U.S. standard, American Society of Mechanical Engineers (ASME) B46.1–2009, contains a definition based on geometrical construction that is difficult to use to obtain an unambiguous result. A number of simplifications in the procedure implemented here were described in Sec. 1, but the procedure can be adapted to more realistic cases.

References

[1] Vorburger TV, Teague EC, Scire FE and Rosberry FW 1979 Measurements of stylus radii, *Wear* **57** 39-49.

[2] Song JF and Vorburger TV 1991 Stylus profiling at high resolution and low force *Appl. Opt.* **30** 42-50.

[3] Song JF and Vorburger TV 1997 Measurement comparison of stylus radii, in *Proc. 1997 Intl. Conf. Prec. Engr.* (ICPE 97, Taipei) **2** 823-828.

[4] ASME B46.1–2009 2010 *Surface Texture (Surface Roughness, Waviness, and Lay),* American Society of Mechanical Engineers, New York).

[5] ISO 5436-2: 2000 *Geometrical Product Specifications (GPS) — Surface texture: Profile method; Measurement standards — Part 1: Material measures* (International Organization for Standardization, Geneva).

Journal of Physics: Conference Series **311** (2011) 012030 doi:10.1088/1742-6596/311/1/012030

Special Properties of Coherence Scanning Interferometers for large Measurement Volumes

W Bauer[1]
Polytec GmbH, Polytec Platz 1-7, 76337 Waldbronn, Germany

E-mail: W.bauer@polytec.de

Abstract. In contrast to many other optical methods the uncertainty of Coherence Scanning Interferometer (CSI) in vertical direction is independent from the field of view. Therefore CSIs are ideal instruments for measuring 3D-profiles of larger areas (36x28mm², e.g.) with high precision. This is of advantage for the determination of form parameters like flatness, parallelism and steps heights within a short time. In addition, using a telecentric beam path allows measurements of deep lying surfaces (<70mm) and the determination of form parameters with large step-heights. The lateral and spatial resolution, however, are reduced. In this presentation different metrological characteristics together with their potential errors are analyzed for large-scale measuring CSIs. Therefore these instruments are ideal tools in quality control for good/bad selections, e.g. . The consequences for the practical use in industry and for standardization are discussed by examples of workpieces of automotive suppliers or from the steel industry.

1. Introduction

Today increasing demands on the functionality of surfaces require precise measurements in the quality control or in R&D with high performance. The demand for ambiguous functionalities of surfaces leads to the compliance of small tolerances. Even tool-marks remaining from the manufacturing process can reduce the function and therefore the surface of the parts must be tested. So for static sealing surfaces, e.g. the flatness is important for the functionality. In other cases two surfaces must be parallel or step heights be manufactured in a very accurate and reproducible way for fittings, e.g. In many cases line profiles of traditionally used tactile systems are not sufficient anymore, because the complete area must be taken into account. On the other hand optical measurement methods deliver fast and reproducible results. A large field of view without time consuming stitching, however, is required for a fast characterisation of a complete area. Therefore coherent scanning interferometers (CSI) –also known as white light interferometers- are an ideal tool, because the performance in z-direction is independent from the field of view [1]. With these instruments 3D-profiles of larger areas (30x30mm², e.g.) can be determined with high precision. In contrast to microscopic set-ups the telecentric beampaths used in large area CSIs allow also measurements of deep lying surfaces without shadowing. The lateral resolution, however, of such a set-up, is reduced. This paper should highlight the chances of optical metrology systems especially for large volumes, but also comparability of the measurements. In the following section facets of practise with CSI are covered based on the experience of users in quality inspection.

2. Coherent Scanning Interferometers for large measurement volumes

2.1. Lateral and spatial resolution of Coherent Scanning Interferometer with telecentric beampath

[1] To whom any correspondence should be addressed.

The most common CSIs use a microscopic set-up. Here the field of view is comparable small and due to the working distance and shadowing effects the characterisation of deep lying surfaces within bore-holes, e.g. is not possible. This is overcome by a set-up with a telecentric beam path with a large field of view. This is typically realized by a Michelson or a Twyman-Green interferometer [2]. In the following investigations a commercially available instrument (TMS-100 TopMap Metro.Lab of Polytec) was used. A measured 3D-profile as an example for the characterisation of a deep laying surface is shown in Figure 1, where the lower surface is 40 mm below the upper surface. As it can be seen the characterised surface is not flat and toolmarks can clearly be identified. Today parallelism of two surfaces, step-heights with a distance of up to 70mm, e.g., or flatness are measured routinely in quality inspection of automotive parts, e.g.

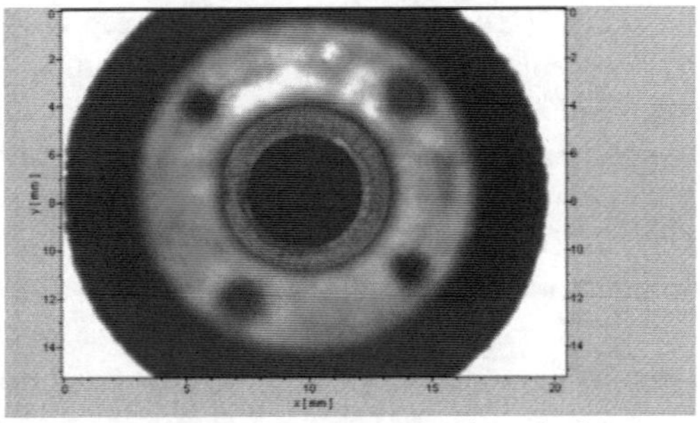

Figure 1. Deep laying surface, the diameter of the work-piece is 19 mm, the lower surface is 40mm below the upper surface

The low aperture of a telecentric set-up, however leads to a lateral resolution, which is poorer as in microscopic systems: According to the Raleigh criteria the resolution R is given by

$$R = 0.61 \times \lambda/NA \qquad (1)$$

Here λ indicates the central wavelength and NA the numerical aperture. Referred to (1) the optical resolution for a telecentric set-up for a field of view of about 30x30mm² is about 39µm -compared to an optical resolution of a microscopic system with a 50x objective, which is about 1µm. The different results of measurements with a telecentric and a microscopic set-up is demonstrated by **Fehler! Verweisquelle konnte nicht gefunden werden.**: These pictures show 3D-profiles of a nebulizer part: On the left-hand side a result is shown obtained with the telecentric set-up of the CSI: Form parameter like flatness, radius or step height can be determined, but small nebulizer holes within the membrane are not resolved. In the 3D profile on the right hand side of **Fehler! Verweisquelle konnte nicht gefunden werden.** determined with the microscopic set-up the small holes and the material throw-out caused by the manufacturing process can clearly be identified.

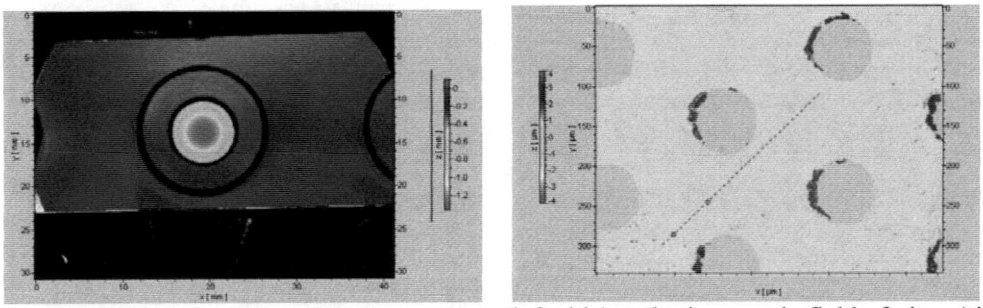

Figure 2. 3D-Profiles of a large area measurement (left side) and microscopic field of view (right side) from the center

The limitation of the optical resolution of CSIs with a telecentric beampath works like an intrinsic low-pass filter. The optical resolution, calculated from the diffraction limit, should not be mixed up with the lateral resolution: Up to now no generally accepted definition exists for the lateral resolution. A possibility is determination by an appropriate standard: We used a depths measurement standard with calibrated rectangular grooves of 90 nm depth and different lattice constants. The rectangularity has advantages for the rectangular adjustment of the standard to the z-axis. The Figure 3 shows the line profile from a measurement with a large area CLI (optical resolution according the Raleigh criteria: 40μm) on a rectangular grating (lattice constant: 800μm, depth: 80 nm). The used instrument has a pixel diameter of 40μm, the same as the calculated optical resolution. At the edges the artefact of so-called batwings can be identified. The batwing effect is caused by superimposition of the interference component of the reference wave and the light reflected or diffracted from the higher surface level on the one hand and the interference between the reference wave and the light propagating back from the lower surface level on the other hand [3]. Because such artefacts lead to wrong values of most GPS-parameter and S-filter has to be applied.

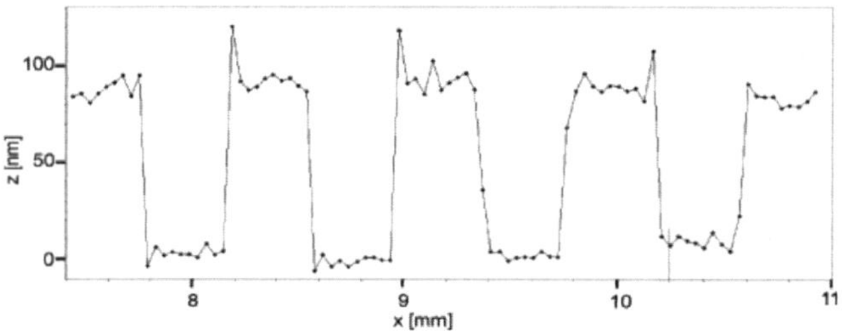

Figure 3. Extracted line profile (original data) of a rectangular grating with a grating constant of 800μm and a depth of 80nm

The part of the grating with smaller lattice constants, however, cannot be resolved completely: The Figure 4 shows the measurement result for the height transfer function for gratings with 5 different lattice constants. A more detailed description can be found in [4], e.g.

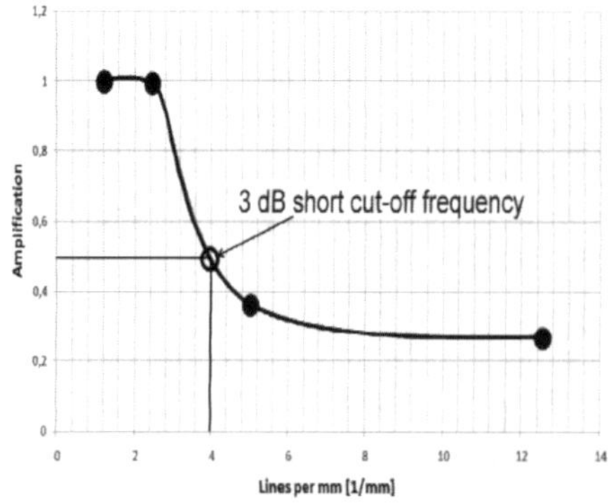

Figure 4. Transfer characteristic of the investigated interferometer with telecentric set-up, when the measurement data is digitally filtered.

From the result shown in Figure 4 a cut-off wavelength of about 250 µm is determined by the 3 dB point. In Figure 5 a measurement is shown on a rectangular grating with a depth of 80 nm and a lattice constant of 80µm, which is twice of the pixel diameter. Therefore the Nyquist Shannon theorem is violated [5]. It is obvious, that the measured profile does not correspond to the real surface structure. So in the case of structures less than twice or even more of the pixel resolution the transfer function according to [6]cannot be defined correctly. Therefore it is also obvious, that roughness parameter with shorter wavelengths than at least twice the pixel resolution cannot be determined with a comparable result to tactile systems with a 5µm radius of the tip, e.g.

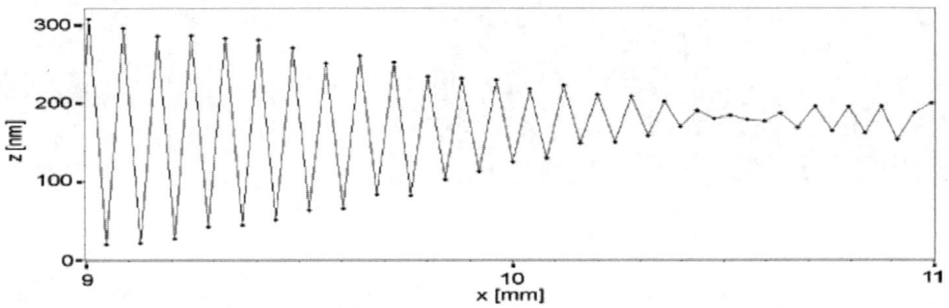

Figure 5. Artefacts appear if a lattice with 80 µm grating constant is measured with a corresponding pixel diameter of 40 µm without filtering

So the spatial resolution is worse than the optical resolution and in order to obtain no artefacts S-filter and low-pass filter have to be applied. The VDI/VDE 2566 part 1.1 guideline indicates as a rule of thumb, that the wavelength of the full modulation depth determined by the 3dB point, is in practice 4 to 6 times greater than the calculated optical resolution [7]. Determined by the 50% modulation depth it is 2-3 times greater. Other authors suppose amplitude decay to 70% for sine structures and 90% for rectangular structures [8]. Although all this is important for the comparability of different instruments and methods the description by common international standards is quite poor, how a user should do this filtering in practise.. A detailed description of possible errors in CSI instruments is given by Gao et al. [9] , who compared microscopic instruments of three suppliers; the VDI/VDE 2566 part 1.3 guideline [10] will also cover telecentric set-ups.

2.2 Special requirements for Coherent Scanning Interferometers with telecentric beampath for the Determination of form parameters

When measuring parameters like flatness, waviness, parallelism results are not affected by the filtering, because here low-pass filter have to be applied anyhow. Step-height determinations are also possible. The evaluation algorithms even for simple flatness measurements, however, are also not standardized: In practise, e.g. a simple spike within the 3D-profile disturbs a flatness measurement . With an a-priory assumption, that a single spike cannot occur, a filter has to be applied to the raw data of a 3D-profile eliminating this spike. By the previously described filtering most of simple spikes are eliminated anyhow. At the moment there is no standard rule, which filter and which parameter an unacquainted user should utilize in the day to day practise: It can be a simple low-pass filter or a median filter with a certain threshold, to name only a few possibilities. The new public-funded German OptAssyst project of research institutes, metrology equipment suppliers and users mainly from the automotive industries works on a user-oriented assistance systems for the reliable use of interferometric and confocal distance sensors in the industrial production.

The fast measurement time of a few seconds make CSIs to an ideal tool for quality control of these parameters, for samples or even for a 100% inspection in the manufacturing line. The need to analyze the complete surface is obvious by the following example: The Figure 6 shows the picture of a sealing surface, on which the outer and inner annular surfaces should be controlled; Taking the complete areas into account the result for flatness is 2.53μm (central picture), whereas two line profiles lead to a result of 1.2μm and 1.6μm, resp., because some of "higher" parts are not hit by the lines.

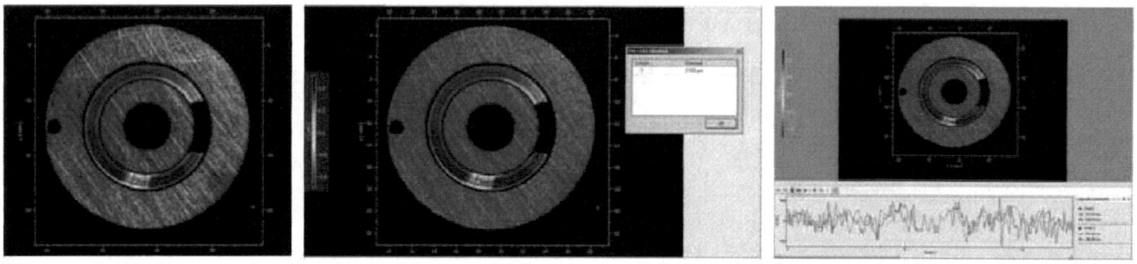

Figure 6. Sealing work-piece, left side: original picture of the work-piece, centre: 3D-profile with the flatness determination, right side: 2 line profiles of the work-piece, location of the line profiles are indicated with dotted lines in the 3D-profile

Another example, where it is very hard to obtain repeatable and reproducible results is shown in Figure 9. The cross section in the picture on the left side shows, that the inner area has the form of a truncated cone. The picture on the right side shows the waviness of a circular-profile of the inner ring. For this part the process capability for quality inspection was not achieved by tactile systems in contrast to the results obtained by CSI-instruments.

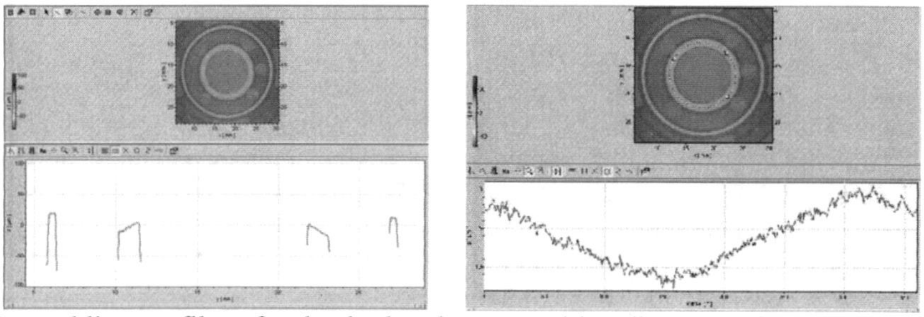

Figure 7. 3D- and line profiles of a shock-absorber part with a diameter of 35mm

In addition measurement time is an important issue for routine measurements in quality control: An example for in-line measurements of piezo fuel injector for car engines is shown in Figure 9 . The vibrations -measured with Laser Doppler Vibrometry- caused by the manufacturing machines are shown on the left side. These vibrations must be suppressed and after that the measured 3D-profile shown on the right side can be obtained.

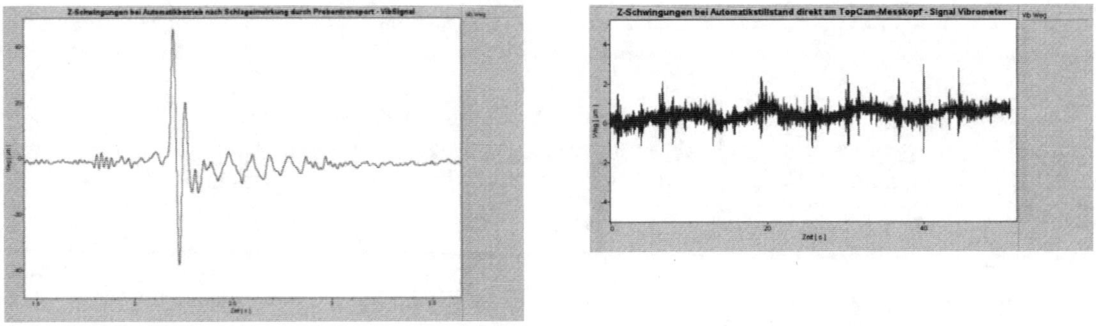

Figure 8. Vibrations after loading (+/- 40µm), relaxation time: about 1s

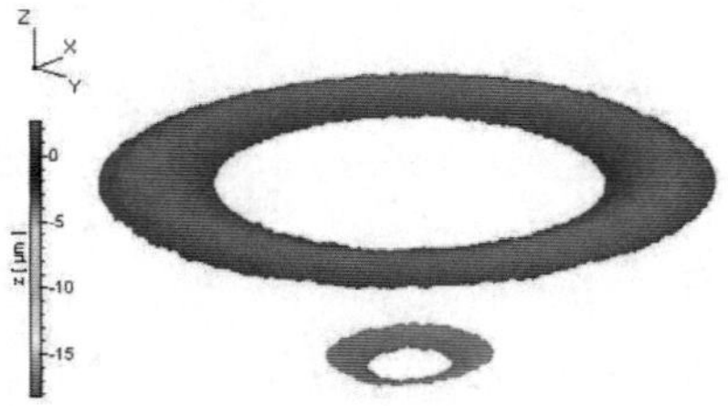

Figure 9. In-Line measurement of a piezo fuel injector. On the left side the vibrations within the manufacturing hall are seen; the displacement is +/- 2µm (y-scal), measurement time (x-scale) is about 50s and on the right side the 3-D Profile after suppressing of these vibrations

The complete cycle for one part includes the loading, the settling down of the vibrations after loading (see Figure 8the automatic finding of the areas of interest, the measurement calculation, export of the result to the data base and unloading. The cycle time is about 6 s. Within this special environment tolerances of flatness and parallelism of about 1 µm can be checked, which means reproducibility and repeatability must be less than 100 nm. Unfortu-nately only few suppliers specify uncertainties of flatness and step-height measurements. Usually only resolution values are given. In most cases this is not very helpful for typical tasks, because here only the minimum resolvable step-height difference is described. In addition, in many cases a statistical value of multiple measurements is published, leading to resolution specifications in the sub-nanometer region. The reason for that might be, that no common rule exists, how to specify flatness or step height measurements for a CSI instrument. We suggest a procedure with a step height artefact measured in the same way as it is done for repeatability and reproducibility determinations in the industry. These specification data, however, give only a first indication, because the real measurement uncertainty depends on the geometry of the sample and the environment. For some workpieces a large z-range is needed to perform the measurement. A higher scan-range can only be covered with a motorized linear stage instead of the more commonly used piezo stage in most CSIs. Motorized linear stages can exhibit relevant positioning errors. Therefore the z-scale of such large-scale CSIs has to be calibrated. As the positioning errors are reproducible they can be calibrated by the help of a correction table [11]. The uncertainty in the vertical direction can be significantly reduced by this calibration procedure. A first analysis of the long-term behaviour of the stages positioning errors by trend shows a good stability. However, the absolute measurement - error is similar also for smaller step heights and therefore an uncertainty value in percent of the step-height is not suitable.

Journal of Physics: Conference Series **311** (2011) 012030 doi:10.1088/1742-6596/311/1/012030

2.3 Using "roughness parameter" calculated from measurements with CSIs for bending classification

Even calculating S_a or S_q roughness parameter in the "traditional way", due to the intrinsic low-path filter caused by the spatial resolution, leads automatically to a waviness window parameter ($W_{a(1\text{-}5)}$ or W_{ca}): The window-parameter $W_{a(1mm\text{-}15mm)}$ allows an accurate classification of critical bending radius (see fig.10).

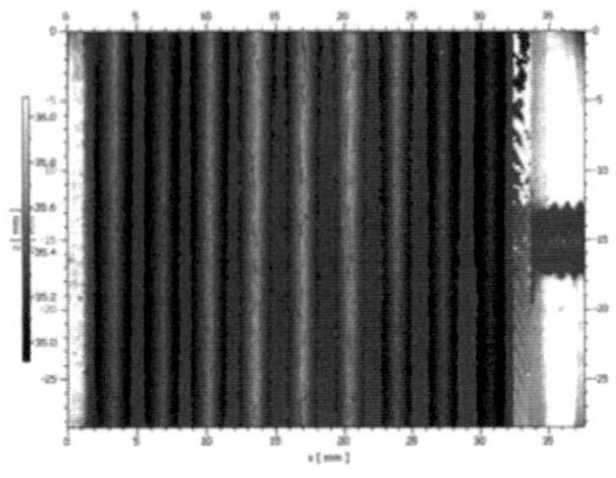

	Sa	Sq
sample 1	8.38 µm	9.82 µm
sample 2	8.66 µm	10.27 µm
sample 3	7.67 µm	9.09 µm
sample 4	9.33 µm	11.76 µm
sample 5	12.48 µm	16.61 µm
sample 6	10.23 µm	12.53 µm
sample 7	7.89 µm	9.81 µm
sample 8	7.83 µm	9.28 µm
sample 9	8.37 µm	9.94 µm

Figure 10. 3D-profile of nine aluminium bending probes and the S_a and S_q values (from left to right)

This allows the qualification of the probes in respect to their bending behaviour. The Figure 11 shows the correlation between the bending radius versus S_a and S_q.

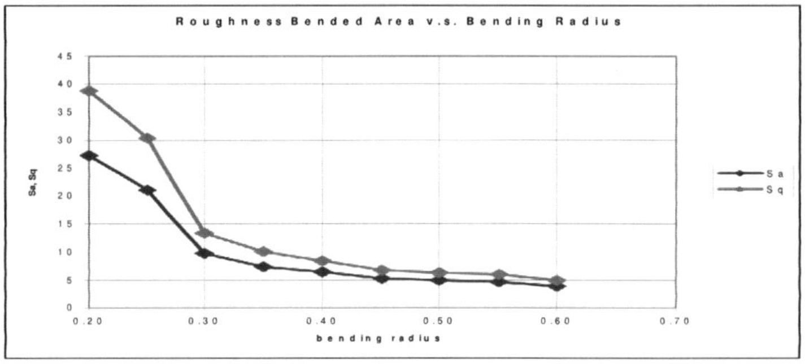

Figure 11. Roughness v.s. bending radius of aluminium (courtesy of W. Hotz, Novelis AG, Sierre (CH))

3. Conclusion

It was shown, that large area measuring CSI is a useful tool for the determination of flatness, parallism and step-heights and in addition surface parameters can be used to determine material properties with high repeatability and reproducibility. Compared to common tactile systems the complete area can be characterized within a short time and therefore ideal for (automated) quality inspection.

References

[1] Dresel T et al. 1992 *Applied Optics* **31** 919.
[2] Malacara D 1992 Twyman-Green Interferometer *Optical Shop Testing,* Ed. D. Malarca, 2nd Ed. New York : John Wiley & Sons
[3] Harasaki A.. Wyant J.C. 2000 *Applied Optics* **39** 2101-2106.
[4] Boedecker S. et al. 2010 Proceedings *Photonic Europe Brussels* **SPIE 7718**-18
[5] Shannon C E 1949 *Proc. of the IRE.* 1949 **37** 10-21.
[6] de Groot P et al. 2005 *Proc. FRINGE,*. s.l. : Springer Verlag, 2006, pp. 30-33.
[7] 2008 *VDI/VDE 2566* **Part 1.1**, Optical measurement and microtopographies, .
[8] to be published *VDI/VDE 2617* **Part 12.2**
[9] Gao F et al. 2008 *Meas. Sci. Technol.* **19**
[10] to be published *VDI/VDE 2566* **part 1.3**: Optical measurement and microtopographies
[11] Boedecker S et al: to be published, **Met&Prop**. **2011**

Journal of Physics: Conference Series **311** (2011) 012031 doi:10.1088/1742-6596/311/1/012031

A minimax fitting algorithm for ultra-precision aspheric surfaces

Xiangchao Zhang, Xiangqian Jiang and Paul J Scott
Centre for Precision Technologies, University of Huddersfield, Queensgate, Huddersfield HD1 3DH, UK

E-mail: x.zhang@hud.ac.uk

Abstract. Aspheric lenses show significant superiority over traditional spherical ones. The peak-to-valley form deviation is an important criterion for surface qualities of optical lenses. The peak-to-valley errors obtained using traditional methods are usually greater than the actual values, which in turn causing unnecessary rejections. In this paper the form errors of aspheric surfaces are evaluated in the sense of minimum zone, i.e. to directly minimize the peak-to-valley deviation of the data points with respect to the nominal surface. A powerful heuristic optimization algorithm, called differential evolution (DE) is adopted. The control parameters are obtained by meta-optimization. Normally the number of data points is very large, which makes the optimization program unacceptably slow. To improve the efficiency, alpha-shapes are employed to decrease the number of data points involved in the DE optimization. Finally numerical examples are presented to validate this minimum zone evaluation method and compare its results with other algorithms.

1. Introduction

Aspheric lenses show notable superiority over conventional spherical lenses in that a multiple-element spherical lens can be replaced by a single aspheric lens. Aspheric surfaces can be represented with [1],

$$z = f(x, y) = \frac{r^2/R}{1+\sqrt{1-(1+k)r^2/R^2}} + A_4 r^4 + A_6 r^6 + \cdots \qquad (1)$$

with $r = \sqrt{x^2 + y^2}$.

Here R is the radius of curvature of the underlying sphere, k is the conic constant and $\{A_i\}$ are the magnitudes of higher order deviations from sphericity.

The form error of a manufactured lens plays an essential role in its performances. Currently the PV, peak-to-valley deviation, is still a very commonly adopted specification for surface quality [2], despite its recognized drawbacks for characterizing surfaces and lack of link to optical performances. Most current commercial software applies the least squares method to fit the nominal surface and calculates the PV error by the difference between the maximum and minimum residuals. However this approach is likely to overestimate the form tolerance and lead to unnecessary rejections.

Here we attempt to directly minimum the peak-to-valley deviation,

$$\min(\max_i d_i - \min_i d_i) \qquad (2)$$

where $d_i = \pm\|\mathbf{R}\mathbf{p}_i + \mathbf{t} - \mathbf{q}_i\|$ is the signed distance from an arbitrary measured point $\mathbf{p}_i$ to its projection $\mathbf{q}_i$ onto the nominal surface. **R** is the optimal rotation matrix and **t** is the translation vector.

This minimax problem is not continuously differentiable, thus very difficult to be solved. This paper presents a heuristic optimization algorithm, called *differential evolution* (DE), to conduct minimum zone evaluation of aspheric surfaces. This method shows great superiorities on stability and accuracy, and makes a good balance between exploration and exploitation.

2. A differential evolution algorithm

At each generation, a Donor vector $\mathbf{v}_i$ is generated for each individual of the population (called *genome* or *chromosome*) $\{\mathbf{y}_i | i=1, \ldots, NP\}$. It is the method of creating this Donor vector that demarcates between the various DE schemes. Two mutation schemes `*DE/rand/1/bin'* and `*DE/current to best/2/bin'* are applied [3,4],

$$\mathbf{v}_i = \begin{cases} \mathbf{y}_r + F(\mathbf{y}_s - \mathbf{y}_t) & \text{rand}[0,1] < p \\ \mathbf{y}_i + F(\mathbf{p}_g - \mathbf{y}_i + \mathbf{y}_r - \mathbf{y}_s) & \text{otherwise} \end{cases} \tag{3}$$

where r, s and t are integers randomly selected from the range [1, NP] (excluding i). $F \in (0,2]$ is used to scale the differential vector, $p \in [0,1]$ is a user-set parameter and rand[0, 1] is a random number uniformly generated in the interval [0, 1].

These two strategies are used very commonly in literature and perform well on problems with distinct characteristics. `DE/rand/1/bin' demonstrates good diversity while `DE/current to best/2/bin' shows good convergence property.

After the mutation phase, a `binominal' crossover operation is applied,

$$u_{ij} = \begin{cases} v_{ij} & \text{if rand}_j[0,1] < CR \text{ or } j = j_{rand} \\ y_{ij} & \text{otherwise} \end{cases} \tag{4}$$

where $CR \in [0,1)$ is a user-specified crossover constant and j_{rand} is a randomly chosen integer in [1, NP] to ensure that the trial vector $\mathbf{u}_i$ will differ from $\mathbf{y}_i$ by at least one component. The subscript j refers to the j-th dimension.

Then a selection operation follows,

$$\mathbf{y}_i^{k+1} = \begin{cases} \mathbf{u}_i^k & \text{if } f(\mathbf{u}_i^k) < f(\mathbf{y}_i^k) \\ \mathbf{y}_i^k & \text{otherwise} \end{cases} \tag{5}$$

with k and k+1 denoting the individuals in the k-th and (k+1)-th generations respectively and f representing the objective function to be minimized. In this paper it is the peak-to-valley deviation.

The pseudo-code for optimization program is shown in Algorithm 1.

```
Input: NP, CR, F, X, y0
// X:data points, y0:initial guess of solution
Initialize population Y based on y0;
k = 0; // generation number
while k <kmax do
   k + +;
   for i = 1 to NP do
      Evaluate fitness f(yi);
      Update global optimum pg;
      Mutation, crossover and selection of yi using Eqs (3)-(5);
   end
   if termination condition satisfied then
      break;
   end
```

end
Output: $\mathbf{p}_g$

Algorithm 1. A differential evolution algorithm

The optimal configuration, i.e. the values of the population size *NP*, the scaling factor *F* and the crossover rate *CR*, is very problem-dependent. According to the 'no-free-lunch theorems', it is not possible to make the optimization program widely applicable whilst maintaining the best performance at every situation [5]. To obtain relatively good performance in different cases, the optimal parameter configuration is individually obtained for each situation. Here meta-optimization is performed off-line using the Local Unimodal Sampling [6]. The running speed of this algorithm is determined by the number of fitness evaluations. Here the number of fitness evaluations is set no greater than 20 000. The unknown variables in the optimization programme are the five motion parameters (Rotation about x and y axes and translation along x, y and z directions), the radius R, the conic constant k and polynomial coefficients $\{A_i\}$ (if applicable). When the shape parameters are all given, which means only the position of the measured data is to be optimized, this leads to a localization problem, and the dimension of the optimization problem will be 5. The recommended parameter settings for different dimensions are listed in Table 1.

Table 1. Parameter setting of DE for aspheric surfaces

D	*NP*	*CR*	*F*
5	23	0.866	0.7549
7	29	0.8745	0.747
8	33	0.8884	0.7347
9	41	0.9046	0.7223
10	45	0.9148	0.7153
11	49	0.925	0.7087
12	52	0.9331	0.6972
13	55	0.9412	0.6808
14	57	0.9542	0.672
15	59	0.9636	0.6631
16	61	0.9729	0.6556

To prevent a too fast decrease of population diversity, the parameters need to satisfy the condition $2F^2 - 2/NP + CR/NP \geq 0$ [7]. Evidently it holds true for all the cases given here.

3. Improving the computational efficiency

This optimization problem is highly nonlinear and lots of local minima exist. To get better results for the DE optimization, the space of the unknown variables is narrowed by supplying a good initial guess for the solution. The orthogonal least squares method [8] is used for this purpose.

3.1. Calculating the orthogonal distances

It is straightforward to write the function of an aspheric surface as $g(\mathbf{x})$=0 with $\mathbf{x}$ being a point on the surface. Normally the projection point $\mathbf{q}$ associated with the point $\mathbf{p}$ is obtained from

$$\begin{cases} g(\mathbf{q}) \\ \nabla g(\mathbf{q}) \times (\mathbf{p}-\mathbf{q}) = 0 \end{cases} \tag{6}$$

using the Gauss-Newton or Levenberg-Marquardt algorithm.

But this method is time consuming, especially when there are many data points. Consequently at the first tens of iterations of the optimization progress, the distance is approximated with [9]

$$d = \pm \|\mathbf{p}-\mathbf{q}\| \approx \frac{g(\mathbf{p})}{\|\nabla g(\mathbf{p})\|} \tag{7}$$

As the motion and shape parameters have been approximately identified using least squares, the distance d will not be very large, i.e. $\mathbf{p}$ is reasonably near to the associated surface. Thus this approximation is acceptable. At the final iterations, the orthogonal distance will be calculated from Equation (6).

3.2. Reducing the number of data points

In practice, the number of measured data points may be up to millions, but actually only dozens of 'significant points' determine the width of the error band (the point number is related with the number of unknown variables). If removing some 'unnecessary' points from the data set, the optimization process can be greatly accelerated. Fortunately, the α-shape technique meets this requirement.

An α-shape is a well-defined polytope, derived from the Delaunay triangulation of a point set, with a parameter $\alpha \in \mathbb{R}$ controlling the desired level of detail [10]. In order to improve the discrepancy, we replace the z coordinates of the data points by the signed residuals resulted from the least squares fitting and then scale the points into a unit cubic. If implementing the 3D Delaunay triangulation, i.e. organizing the discrete points into a set of tetrahedra, the real 'key points' are likely located at some tetrahedra with large circumscribed spheres, thus some tetrahedra with small circumscribed spheres can be omitted. But many 'boundary points' will be retained unnecessarily, hence the boundary and interior points are handled separately. Viewing from the z direction, the boundary points are recognized using the modified Graham scan method [11]. In the programme, the tetrahedra are sorted by their radii of circumscribed spheres (in descending order). The set for the points to be kept is initialized as null, and then the vertices of each tetrahedron are checked successively. The checking procedure is presented in Algorithm 2.

```
Input: point set {x_i}_{i=1}^N, points to be retained Y = ∅ 4, Delaunay tetrahedra T ∈ N^{M×4};
Find the boundary points B;
for j = 1 to M do
   for i = 1 to 4 do
      if T_ji ∈ B then
         if T_ji has the greatest positive or the smallest negative z coordinate among the four vertices of T_j then
            Put T_ji into Y;
         end
      else
         Put T_ji into Y;
      end
   end
   if The number of points in Y satisfies the pre-set limit then
      break;
   end
end
Output: Y;
```

Algorithm 2. The checking process for the vertices to be retained

Figure 1 shows a 2D example. Given 300 points, 30 points on the envelop are sampled using the α-shape technique. The number of points to be retained is directly related with the number of unknown variables. For example, in the evaluation of flatness (resp. sphericity), there are three (resp. four) variables and four (resp. five) extreme points are needed to calculate the minimum zone error. For the same reason, if n polynomial terms are involved in the aspheric function, at least n+8 points are needed. To avoid removing extreme points by mistake, $2n$+14 points will be kept in the DE optimization.

(a) Given point set (b) Envelop points

Figure 1. Point reduction using α-shapes

When a set of significant points are selected, DE optimization is undertaken. Then the residuals of all the data points with respect to the fitted surface are calculated. If some data points are not contained within the tolerance zone, these exterior points are added into the significant point set and the minimum zone evaluation program is run again. This procedure is repeated until all the data points are contained within the tolerance zone.

4. Experimental validation and discussion

The validity of the proposed algorithm is verified with a data set of 8100 points, as shown in Figure 2. The unit of length here is mm if not specified otherwise. The nominal shape parameters are given as R=520, k=-0.7, A_4=5.2e-5, A_6=-6.5e-6, A_8=3.11e-8, A_{10}=3.222e-9. Noise with its amplitude σ=1μm is introduced using the fractal Brownian function [12]. The dimensionality of the optimization problem is 11 and the control parameters in the DE program is adopted as NP=49, CR=0.925 and F=0.7087, in accordance with Table 1.

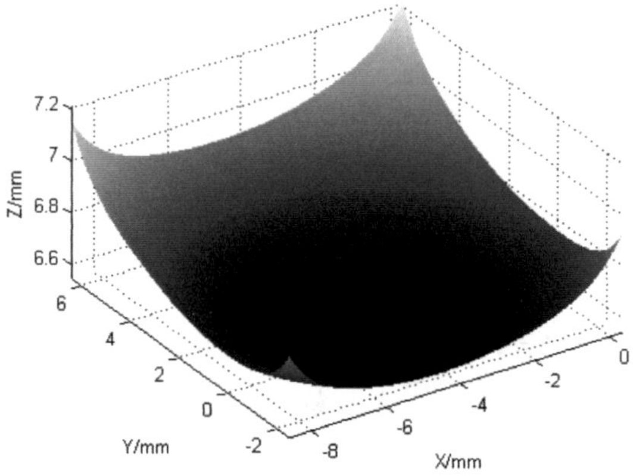

Figure 2. An aspheric surface

The Differential Evolution optimization program was coded in MATLAB R2009a and run on a PC with Intel(R) Core (TM) 2 Duo CPU8500 3.16GHz, 3.24GB RAM. The program was run 100 times and the obtained PV errors and running time (including ODF least squares initial fitting, alpha-shape point reduction and DE parameter optimization) are listed in Table 2. For comparison, the direct DE optimization without point reduction was run 100 times and the corresponding results are also presented.

Table 2. Fitted results of the aspheric surface

method	LS ODF	DE		
		max	min	mean
PV/μm	3.79	3.28	3.21	3.22
		with α shape		without α shape
Time/s	0.14	5.40		194.93

In the table it can be seen that the proposed DE optimization can obtain a much smaller PV form error compared with the least squares ODF, meanwhile, at a cost of much longer running time. But the average time for the whole fitting process is less than six seconds, which is acceptable in practical applications. The point reduction technique using α-shapes reduced the running time by 97%.

5. Conclusions

This paper proposes an optimization method using differential evolution to evaluate the peak-to-valley form errors of aspheric surfaces. To make the optimization program specifically works well for different dimensions of unknown variables, the parameter configuration is obtained using meta-optimization. Additionally, the alpha shapes is adopted to get the 'envelop points' which potentially determine the PV form error, so that the points involved in the optimization program are greatly reduced. Experimental results prove the running time can be reduced by 97% without influencing the optimization results. This program can also be utilized in aspheric surface fitting to calculate the shape parameters from the given measured data points or surface matching to determine the optimal relative position between the measured data and the nominal surface.

Acknowledgments

The authors gratefully acknowledge the European Research Council for its programme ERC-2008-AdG 228117-Surfund.

References

[1] ISO 10110-12:2007 Optics and Photonics-Preparation of Drawings for Optical Elements and Systems-Part 12: Aspheric Surfaces
[2] ISO 10110-5:2007 Optics and Photonics-Preparation of Drawings for Optical Elements and Systems-Part 5:Surface Form Tolerances
[3] Price K and Storn R 1995 Differential evolution- a simple and efficient adaptive scheme for global optimization over continuous spaces, *Technical Report*, Intern. Computer Science Inst., Berkley
[4] Qin A K and Suganthan P N 2005 Self-adaptive differential evolution algorithm for numerical optimization, in: *2005 IEEE Congress on Evolutionary Computation* **2** 1785–91
[5] Wolpert D H and Macready W G 1997 No free lunch theorems for optimization, *IEEE Trans. Evol. Comput* **1** 67–82
[6] Pedersen S M E H 2010 Tuning and Simplifying Heuristical Optimization, *PhD thesis*, University of Southampton, UK
[7] Zaharie D 2002 Critical values for the control parameters of differential evolution algorithms, in: R. Matoušk, P. Ošmera (Eds.), *Proc. 8th Intern. Conf. Soft Computing Mendel*, Brno, Czech Republic 62–7
[8] Zhang X 2009 Freeform Surface Fitting for Precision Coordinate Metrology, *PhD thesis*, University of Huddersfield, Huddersfield, UK
[9] Taubin G 1991 Estimation of planar curves, surfaces and nonplanar spaces curves defined by implicit equations with applications to edge and range image segmentation, *IEEE Trans. Patt. Anal. Mach. Intell.* **13** 1115–38
[10] Edelsbrunner H and Mücke E P 1994 Three-dimensional alpha shapes, *ACM Trans. on Graphics* **13** 43–72
[11] Zhang X, Jiang X and Scott P J 2008 Orthogonal distance fitting of precision free-form surfaces based on l_1 norm, in: F. Pavese, M. Bär, A. B. Forbes, et al. (Eds.), *Adv. Math. and Comput. Tools in Metrology and Testing VIII*, World Scientific 385–90
[12] Mandelbrot B B and van Ness J W 1968 Fractional brownian motions, fractional noises and applications, *SIAM Review* **10** 422–37

AUTHOR INDEX